How to Read and Interpret
SCHEMATIC
DIAGRAMS

How to Read and Interpret
SCHEMATIC DIAGRAMS

J. RICHARD JOHNSON

HAYDEN BOOK COMPANY, INC.
Rochelle Park, New Jersey

Library of Congress Cataloging in Publication Data

Johnson, J. Richard
  How to read and interpret schematic diagrams.

  Includes index.
  1. Electronics — Diagrams. I. Title.
TK7866.J63 1982   621.381′.022′3   82-11924
ISBN 0-8104-0868-6

2	3	4	5	6	7	8	9	PRINTING
83	84	85	86	87	88	89	90	YEAR

Preface

The wide use of schematic diagrams in every phase of technology has made a good working knowledge of them very important to most who are interested in the function and application of scientific devices—particularly in the field of electronics. The student of electronics must develop an understanding of schematic and block diagrams as he learns about electrical principles. The same is true of the practical technician, serviceman, and experimenter.

The purpose of this book is to teach the meaning and use of the schematic diagram in a systematic manner, so the principles may be more easily learned in a step-by-step rather than random manner.

The treatment in this book starts with explanations of the meaning and purpose of schematic diagrams, block diagrams, and wiring diagrams. With this groundwork established, it proceeds to examples of diagrams of various types of electronic equipment. The book develops in a step-by-step manner: introductory information in Chap. 1, schematic symbols and basic building blocks in Chap. 2, synthesis into diagrams in Chap. 3, and examples of actual equipment diagrams in the remaining chapters.

The examples of equipment diagrams in Chaps. 4 through 9 are for general illustrative purposes. To show how diagramming is accomplished in these examples, it is necessary in many places to explain principles of operation. No attempt is made to include complete circuit analyses for which much more space would be necessary. Rather, the purpose is to give the reader examples which will familiarize him with the techniques of drawing and interpreting the diagrams he encounters. It is expected that this will enable him to apply the same reasoning to many other types of equipment not covered here.

For the above reason, it is expected that this book will not be a substitute for a study of basic electronic principles, enabling the reader to master the principles of operation of the devices discussed. But, it is a complement to a study of those principles, enabling him to read and interpret schematic diagrams and make use of them in testing, repair, and design.

I am very grateful for the efforts of Ella Johnson for typing most of the first draft, to Nancy Scholz for most of the final typing, and to Barbara Brown for much intermediate typing, editorial work, and helpful encouragement.

Special mention is in order here, in acknowledgment of RCA's kind permission to reproduce many illustrations from their book *Solid State Devices Manual.*

J. RICHARD JOHNSON

Contents

CHAPTER 8:

Interpreting Television Receiver Diagrams   154

CHAPTER 9:

Computer Diagrams   178

Index   191

1

The What and Why of the Schematic Diagram

Imagine a TV repairman working in a shop to repair a receiver. What does he have beside him for constant reference? A schematic diagram.

Now consider a military technician checking out a piece of military electronics before it is sent out to the field. What does he use as a reference? A schematic diagram.

An engineer is developing a new idea or piece of equipment. As he records his idea, what is the most likely form for it to appear in on paper? A schematic diagram.

A radio amateur builds a new piece of equipment or changes the wiring in his present equipment. His most important reference—the schematic diagram.

Certainly by now (if you didn't long ago) you realize that one of the most important phases of any work with electronic or electrical equipment is the schematic diagram. If you expect to do any of the things mentioned above, or have anything to do with handling electrical or electronic equipment, you really need a knowledge of and skill with such diagrams. You will find also that certain other types of diagrams, particularly the block diagram, are closely related to the schematic type. For this reason, we shall also give some attention to these other types, so that schematic diagrams can be better understood. We start with a definition of the schematic diagram, before we explain how it is used.

Definition

What is a schematic diagram? We should answer this question before we use one. *Schematic* is developed from the word *scheme*. Thus, the schematic diagram shows the scheme of things. Its major message is the purpose of the equipment and the scheme by which it achieves its purpose. Therefore, it must present all information necessary to tell the user how the piece of equipment functions electrically. (There are also mechanical schematics, which show

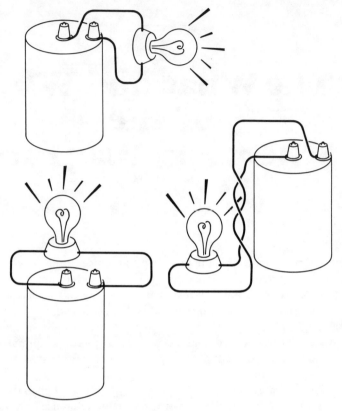

Fig. 1-1   *Several physical arrangements of a lamp connected to a battery, all electrically the same.*

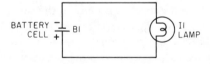

Fig. 1-2   *The schematic diagram for any of the arrangements of Fig. 1-1.*

mechanical function.) It shows what parts are used and how they are connected together. It does so from a purely electrical point of view and its presentation is never hampered by attention to the physical or mechanical design of the equipment. For example, a connection appears the same on the schematic diagram whether the connecting wire is only a quarter of an inch or 20 feet long.

Purpose

The purpose of a schematic diagram is implicit in its definition. It is prepared to provide a graphic analysis of the necessary electrical components of a piece of equipment, and the manner in which they are electrically connected to function as a unit. However, the diagram is not an *explanation* of the function; the user must be able to *interpret* the diagram. The purpose of this book is to help you acquire the ability to interpret!

Related Types of Diagrams

The schematic diagram should not be confused with other types of diagrams often used with it. To minimize the possibility of such confusion, we consider in this book such other types of diagrams as functional flow diagrams, block diagrams, wiring diagrams, and physical layout diagrams. These all differ from the schematic diagram in that their prime purpose is something other than showing just the functional idea of the circuits involved.

A Simple Example

Consider the circuit variations shown in Fig. 1-1. The schematic diagram for all these is the same; this diagram is shown in Fig. 1-2.

Two component symbols are used. These are of the type used in schematic diagrams; there is a symbol for each of the components in any schematic diagram. Notice that the battery cell is simply a long thin line beside a shorter thicker line. The lamp is a line forming a loop of particular shape. In Chap. 2, we shall discuss various component symbols and how they are combined into circuits.

Why is the schematic diagram exactly the same for circuits that have such a different appearance? The answer is: the schematic diagram is made to show only the basic electrical functioning of the circuit, and not any of its physical characteristics. From this standpoint, the circuit arrangements of Fig. 1-1 are all the same. Negative electric charges flow from the negative terminal of the battery, through the lamp, and back to the positive terminal of the battery. This is true regardless of where the lamp is located and how the conductors are arranged (short, long, twisted, etc.) as long as they connect the negative terminal of the battery to one side of the lamp and the positive terminal to the other side.

Schematic Does Not Show Physical Features

Here is a very important rule about schematic diagrams:

Rule 1. *The arrangement of conductors and components in a schematic diagram bears no necessary relation to their actual physical arrangement.*

You will notice that we say, "necessary relation."

There is nothing to say that we are barred from making physical arrangements of schematic diagrams conform in some ways to the actual physical features of the equipment depicted. This is not often done, however, because such conformance might interfere with the primary purpose of the diagram, which is to show circuit function.

To recap: The purpose of a schematic diagram is to show, in the simplest, most useful form, the components making up a circuit, how they are interconnected, and how the circuit functions.

The names of the components and the way they are interconnected are perceivable to almost anyone. Understanding of the functioning, of course, requires that the user have some knowledge of circuits in general and can quickly recognize what types of circuits are used and how they function. He is then able to use the diagram to analyze, troubleshoot, repair, or even just operate the equipment better.

Connections Are Straight Lines

Meanwhile, let us examine the diagram of Fig. 1-2 further. Notice that the connections between the lamp and battery are straight lines, even though in the actual case the conductors are not straight. Why? Because the purpose of a schematic diagram is to show electrical function. The diagram's message as to function is clearest if the diagram is neat, and it is almost always neatest when straight lines are used.

It should be kept in mind that a straight thin line represents a connecting conductor *without regard to the physical nature of the conductor.* The conductor could be a hair-thin wire or a huge bus bar a foot wide and an inch thick, and it would still appear only as the same thin straight line on the schematic diagram. Usually the lines on a given diagram represent at least several different types of conductors as the functioning of the circuit requires, but they look the same on the diagram.

Merging and Crossing Connections

Sometimes several conductors merge, that is, connect together, in a circuit. This is shown by placing a dot at the point at which the conductors cross, or where one conductor terminates as it reaches another, as illustrated in Fig. 1-3(A). Notice that, where conductors cross, but *do not* connect, either the dot is simply omitted (B) or a jumper is formed to indicate that one conductor

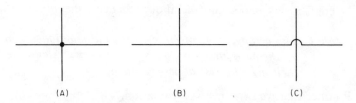

(A)   (B)   (C)

Fig. 1-3 *Symbols for conductor crossovers on schematic diagrams: (A) conductors connected, and (B) and (C) conductors not connected.*

"jumps over" the other to avoid connecting to it [Fig. 1-3(C)]. The jumper symbol is not often used today, and you can expect to encounter mostly the undotted crossover.

Sometimes a heavier line (Fig. 1-4) is used in a schematic diagram to indicate that the "conductor" so depicted is either a heavy ground *bus*, a metal strip or the metal chassis or frame of the device. If it is grounded, the symbol for ground or chassis ground is used as illustrated in Fig. 1-4.

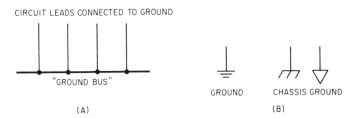

CIRCUIT LEADS CONNECTED TO GROUND

"GROUND BUS"

(A)

GROUND   CHASSIS GROUND

(B)

Fig. 1-4 *(A) Ground bus symbol with leads connected, and (B) symbols for single-ground connections.*

Grounds

The concept of ground should really be defined. The origin of the term is clearly related to connections made to the earth itself. Early telegraph systems made use of earth as a *return* lead, so that only one wire was necessary. Electric power distribution systems have always connected one terminal of the power source to earth. Today, the main purpose of this is safety; it prevents large metal frames in or on the ground from accidentally becoming charged and dangerous.

The term *ground* has now been expanded to include any common connection to which a point in each circuit section of a chassis or module can be connected without causing interaction or interference among these circuit sections. This common connection can be either connected or not connected to earth ground.

We have established in this chapter a preliminary idea of what a schematic diagram is. It should be noted that each component has its own symbol. We shall see that these symbols are laid out in a schematic diagram in such a way as not only to show how they are connected, but also how they contribute to function and functional flow in the device depicted.

You have probably also noticed that, for each kind of graphic schematic symbol, there is a letter symbol, most often followed by a number. These letter designations are used to identify specific components in a separate parts list. This is discussed further in Chaps. 2 and 3.

2
Component Symbols and Diagram Formation

Chapter 1 describes the nature and purpose of a schematic diagram, and the philosophy of its use. In this chapter we discuss the symbols found in such diagrams and something about how they are combined to form the complete diagram of a circuit section, a device, or a whole piece of equipment.

Connections

We discussed something of the nature of connections in Chap. 1. It should be kept in mind that the schematic diagram is not designed to give any information about the physical nature of a connection, that is, whether the connection is made by large or small wire, whether it is long or short, etc. The diagram shows only whether or not a given connection is made; if there is a straight line between two points, they are connected; if there is no line they are not connected. Whether a connecting wire is many feet long, or the terminals of two components are directly soldered together, the appearance on the schematic diagram is simply a straight line of diagrammatically convenient length.

Functionally related symbols are placed near each other and then the connecting lines are drawn appropriately between them, with their paths determined by neatness and simplicity. It often happens that the components that are grouped in one part of a schematic diagram are also near each other physically in the equipment. But *this need not be*, and frequently is not the case.

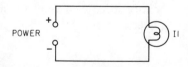

Fig. 2-1 Sometimes the power source itself is not shown on a schematic diagram; instead, points of power application are shown by terminals as in this diagram.

6

Sometimes components whose functions are closely related are located in different parts of the equipment for convenience. For example, the on–off switch of a device must be located on the front panel (or otherwise convenient to the user). However, its function is to connect or disconnect the power supply which may be located in a different part of the unit. But, on the schematic diagram the symbols would be located close together.

Power Sources

In Fig. 1-2 is the simplest type of schematic diagram. It depicts the connection of a power source (the battery) and a user of power (the lamp). A majority of the circuits in schematic diagrams require at least one source of power. In the case of Fig. 1-2, the power source is completely defined by the battery symbol. In many diagrams the actual power source is not diagrammed, and there are merely terminals through which power is brought in. In this case, the terminals are assumed to be connected externally to the power source. This is illustrated in Fig. 2-1.

A terminal is a screw, binding post, or lug. In this case, we are talking about such a device at the edge of a circuit board or chassis, or the outer wall of a box or cabinet. Its purpose is to connect some point in the circuit to something outside the circuit or its physical mounting. Of course, terminals are also used within circuits.

Batteries

For discussions of the basic principles of a circuit, the power source is often shcwn as a battery, even if this will not be the type of source finally used.

Consider the battery symbol used in Fig. 1-2. Notice that the "battery" in this case is only one cell, indicated by the combination of a short, thick line (representing the negative terminal) and a longer, thinner line (representing positive). This pair forms a *cell symbol.* Strictly speaking, a single cell should not be referred to as a *battery,* the term is reserved for combinations of cells connected together. To provide more voltage than can be supplied by one cell, two or more cells can be connected together in series to form a battery, as shown in Fig. 2-2. At the left, the two cells are shown as physically separate entities.

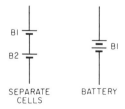

Fig. **2-2** *Drawings showing how a distinction can be made on the schematic diagram between separate cells connected together and a battery of cells.*

Fig. 2-3   *Examples of available batteries, each containing two or more cells connected internally.*

However, most batteries have the cells combined into one physical structure, whose diagram symbol is shown at the right in Fig. 2-2. The physical appearance of typical combined units is shown in Fig. 2-3.

Letter Symbols

We should now give attention to another very important feature of the schematic diagram, namely *letter* symbols. In most cases, the user of the diagram must be able to identify and designate any particular component on the diagram. Each type of component has its own letter symbol; the letter symbol is then followed by a number to indicate which of the components on the diagram having the same letter symbol is being designated. For example, in Fig. 2-1, the lamp is designated I1. The *I* is the letter symbol for any lamp. The *1* indicates that this is the first lamp designated in this diagram. If there were another lamp, it would be designated *I2*, and successive lamps encountered would be *I3*, *I4*, etc. As we shall see, these designations are used in parts lists which accompany schematic diagrams. (Parts lists allow the user to determine the specifications of any particular component in the diagram.) As the graphic symbols for the various components are introduced in this chapter, the letter symbols will also be presented. It is important that these be learned.

The distinction between physically combined and separate cells is often not made on a schematic diagram. If it is, it is done as shown at the left in Fig. 2-2. The distinction is ordinarily made only when there is some important relation between it and the function of the circuit. Usually the applicable rule is as follows:

> **Rule 2.** *Such features as the number of cells in a battery, the number of plates in a capacitor, and number of turns in a coil are never dependably indicated by anything in the symbol. Such information is normally supplied by a label near the component, or by a reference in the parts list (keyed by a symbol number: B1, R2, C3, etc. Parts lists and reference codes are discussed later).*

Thus, the symbols for a battery and a lamp, shown in Fig. 1-2, tell us nothing about the voltage applied to the lamp, or the lamp's power and current rating. The diagram serves its purpose by telling us that a battery is connected to a lamp. In fact, as mentioned earlier, sometimes the battery symbol may stand for *any* power source and not necessarily just a battery. The latter is true, often, for simple diagrams used in the early stages of the development of an idea, or when simplification of the principle of a larger circuit diagram is desired. Large, complete, and "official" diagrams usually indicate the specific type of power source used.

Let us consider other sources of power. One of the most common is a rotary generator or motor-generator.

Rotary Generator Sources

This type of source can provide either direct or alternating current, produced by the motion of the rotating coils through the magnetic field produced by the field magnet. Generators that produce alternating current are usually referred to as *alternators*.

Alternators can have any of several arrangements:

1. Coils rotating in a magnetic field produced by a permanent magnet generate the voltage output.

2. Coils rotating in a magnetic field produced by an electromagnet in which the (field) coil is energized from a separate source.

3. The rotating structure is either a permanent or electromagnet and its field rotates through the fixed coils in which the output voltage is generated. If the rotor is an electromagnet, it must carry what is known as *exciting current*.

For generators of any size, method three is virtually always used because the high-voltage and current outputs are easier to handle from a fixed structure.

To derive output in alternator types 1 and 2 and to feed in the exciting current in type 3, some means is needed to connect the rotating coils to an outside source of current for excitation. For ac generators this means is in the form of *slip rings*. These are two continuous rings of metal, insulated from each other and mounted side by side on the rotating shaft, and connected to the ends of the rotating coil. Contacting these are two fixed *brushes* (usually of carbon) which connect to the external circuit. The symbols most commonly used for rotary ac generators are shown in Fig. 2-4.

Sometimes it is necessary or desirable to show whether the generator is of the permanent magnet (PM) type or if a field coil (or coils) is used and whether it is connected as a series, parallel, or compound *wound* generator. In that case, the field coil(s) may be shown separately, as illustrated in Fig. 2-5.

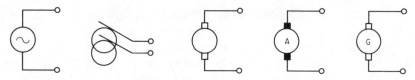

Fig. 2-4   *Basic generator symbols.*

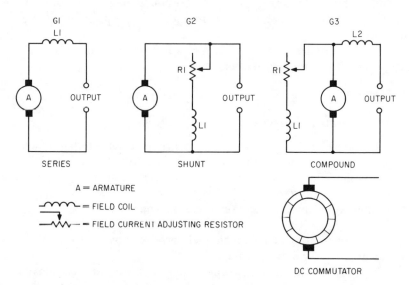

Fig. 2-5   *Schematic symbols showing field-coil connections in three types of basic rotary generators and the use of a dc commutator.*

For dc generators, the same symbols can be used. Instead of slip rings, the dc generator uses a commutator so that the current continues in one direction, instead of reversing in direction each half-cycle. The last part of Fig. 2-5 shows how the commutator is sometimes indicated in a symbol for a dc generator.

Graphic symbols for generators are often used without letter symbol designators, because in such cases the only needed information is given in a label. When a letter symbol is used, it is usually a *G*.

Generator symbols are not often used in schematic diagrams of electronic equipment because such equipment usually derives its energy from a *power supply.* Actually such a device is not an ultimate source, but a device that can modify the output of other sources to suit the needs of the equipment with which it is used. For example, most common is the power supply that connects to the commercial mains (house wiring), converting an ac voltage to one or more dc voltages needed by the equipment. It is deemed sufficient in such a case to diagram the power supply back to the point at which it plugs into the power mains. Other types of power supplies connect to a battery (such as the one in an automobile) and convert the battery dc voltage to other dc voltages as required. These devices are most often called *inverters*.

The power supply as a source is not just a component, but a full circuit of its own. Common circuit diagrams are discussed in Chap. 4.

Resistors

The resistor is probably the most common component used in electronic equipment. It appears in any of a number of different physical forms, but the schematic symbols are of three basic types: fixed, fixed-tapped, and variable. Resistors typical of these three forms are shown in Fig. 2-6, along with the corresponding schematic and letter symbols. There are many variations in the physical construction of resistors. Since we are interested here primarily in the schematic symbols, we shall not take space to discuss construction details.

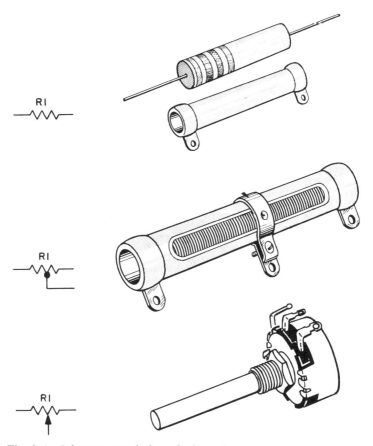

Fig. 2-6 *Schematic symbols and physical appearance of three types of resistors.*

However, it should be said that many resistors today are not discreet components such as are shown in Fig. 2-6, but are part of tiny deposits on substrates of integrated circuits. These can be of either the fixed or fixed-tapped variety. The letter symbol is always *R*.

The connections of the resistors are indicated by the schematic diagram. Also very important is the value of the resistance in ohms. The ohm is the amount of resistance required to cause a voltage drop of 1 volt across it with a current of 1 ampere through it. Resistors are available with resistance from a fraction of an ohm to many megohms (millions of ohms). The value in ohms of a resistor is usually given in either of two places: in the parts list accompanying the schematic diagram, or from a label on the schematic diagram alongside the symbol of the resistor.

The resistance value of a composition (carbon) type tubular resistor is indicated on the body of the resistor by rings of color which relate to a standard code. This code is defined in Fig. 2-7, and an example of how it works is given in Fig. 2-8. The code is an important key in correlating a resistor symbol on a schematic diagram with a resistor in the equipment.

COLOR	SIGNIFICANT FIGURE	MULTIPLYING VALUE	TOLERANCE (%)
BLACK	0	1	± –
BROWN	1	10	± 1
RED	2	100	± 2
ORANGE	3	1,000	± 3
YELLOW	4	10,000	± 4
GREEN	5	100,000	± 5
BLUE	6	1,000,000	± 6
VIOLET	7	10,000,000	± 7
GREY	8	100,000,000	± 8
WHITE	9	1,000,000,000	± 9
GOLD	–	0.1	± 5
SILVER	–	0.01	± 10
NO COLOR	–	–	± 20

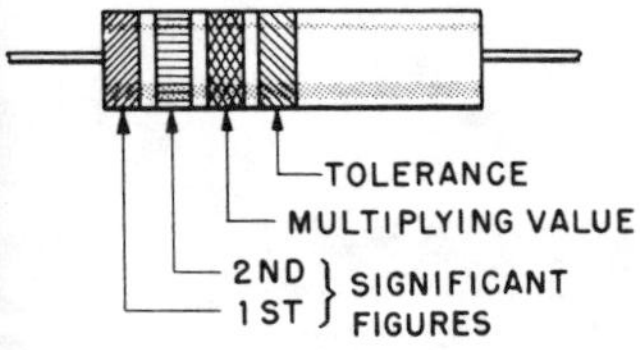

TOLERANCE
MULTIPLYING VALUE
2ND } SIGNIFICANT
1ST } FIGURES

Fig. 2-7   Resistor color code.

Capacitors

Although resistors are probably the most common of electronic components, capacitors undoubtedly come next. Essentially, a capacitor comprises two flat metallic surfaces separated by insulating material. The surfaces are called *plates* and the insulating material the *dielectric*.

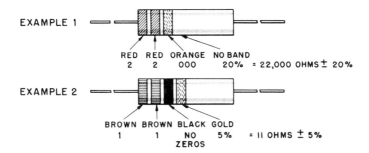

EXAMPLE 1

RED   RED   ORANGE   NO BAND
 2   2   000   20%   = 22,000 OHMS ± 20%

EXAMPLE 2

BROWN BROWN BLACK GOLD
  1   1   NO   5%   = 11 OHMS ± 5%
               ZEROS

Fig. 2-8   *Example of the use of the resistor color code.*

Examples of types of capacitors and their schematic and letter symbols are shown in Fig. 2-9. Fixed capacitors come in a wide variety of constructional forms. The simple fixed type in Fig. 2-9 (A) may have a dielectric of paper, mica, mylar, or other materials. One special type is the electrolytic capacitor in Fig. 2-9(B). It is used where large capacitances (one microfarad to thousands of microfarads) are required and where low-loss high-frequency characteristics are not necessary. Electrolytic capacitors are polarized; one plate must be operated at a positive dc potential above the other plate. Since it is thus necessary to know which terminal is which, electrolytic capacitors are represented by the polarized capacitor symbol of Fig. 2-9(B). All capacitors have *C* as a letter symbol.

Variable capacitors are those in which the position of one or both plates can be varied to change capacitance by a change of spacing and/or overlap of the plates. Examples of the symbols are shown in Fig. 2-9. The most familiar is the type in which the plates are two sets of metal sheets, interleaved, mounted so that one set can rotate into or out of mesh with the other set. The assembly of the rotating plates is called the *rotor* and the fixed set the *stator*.

Another type of variable capacitor is called a *compression trimmer*. In this type there is a fixed flat plate, usually mounted or deposited on a ceramic base, and a spring-like variable plate. The position of the variable plate above the fixed plate can be varied by rotation of a screw, which pushes the variable plate closer to the fixed plate or allows it to spring farther away. The trimmer may use air for the dielectric, or may have a layer of mica between the plates. The symbols for either type are shown in Fig. 2-9(C).

It is sometimes desirable to use two rotary variable capacitors so that they rotate together. The two major arrangements are *split stator* and *ganged variable* setups. In either case, the two capacitors are often built physically into a ganged arrangement, that is, with the two stators insulated and the rotors on a common shaft. Symbols for these arrangements are shown in Fig. 2-9(D) and (E).

In the case of the split-stator capacitor, two sections are used. It is particularly useful in balanced circuits in which the common rotor needs to be

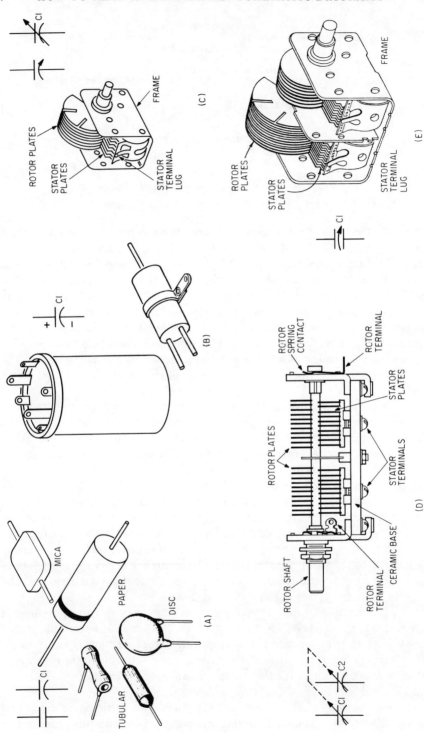

Fig. 2-9 Examples of the types of capacitors and their schematic symbols.

grounded and the two stator sections provide balance on each side of ground. The most common application is in tank circuits, discussed later.

Ganged capacitors are used where several different circuits must be tuned in synchronism or with a fixed frequency difference between them. The most common application is radio receivers, in which at least the mixer input and oscillator sections must be tuned together, so that they *track*; that is, so that the frequency of the oscillator is always separated from the frequency of the incoming signal by an amount equal to the intermediate frequency. Ganged capacitors are not limited to two sections (as are the split-stator types); for example, if a superheterodyne radio receiver has a tuned RF amplifier, it needs three sections (one each for the RF amplifier, the mixer, and the oscillator). As many as five sections have been used in special applications.

Inductors

Inductors, or coils, as they are more commonly known, exist in a number of forms in electronic and electrical equipment. The more common forms, and the variations in their schematic symbols, are shown in Fig. 2-10. Note that the letter symbol is *L*.

The first form illustrated is the simple untapped air-core coil. As can be seen in the figure, these come in a number of forms, all having the same schematic symbol. The simplest is the self-supporting air-core coil in which the stiffness of the wire, formed into the desired helical shape, supports the coil. This type is limited to a small number of turns but is desirable for high radio frequency operation because of its relatively low losses. Somewhat more common is the same kind of winding wound over a cylindrical form. The form allows more turns to be used, because of the physical support the form provides. The form does add some losses at high frequencies, however, even when the material is carefully selected. A single-layer helical coil, with or without a form, is often referred to as a *solenoid winding*.

Coils with multiple-layer windings are also frequently used. A popular multilayer type is the *universal* winding. This may appear in the form of a single winding, or with two or more windings alongside each other on the same form. In the latter case, the windings are called *pi* windings and the overall assembly is called a *pi-wound* coil. Universal windings are popular for low-frequency RF coils, and for IF coils and RF choke coils.

Sometimes two or more simple layers are used to build up many turns in a limited space. High-frequency losses are greater in this type than in the universal winding. This is because the wire from the end of the first layer is wound back across the first layer, bringing turns from different parts of the coil close together. For this reason, such windings are most popular in power transformers and filter chokes. In all of the above cases, the appropriate schematic symbol is the one in Fig. 2-10(A).

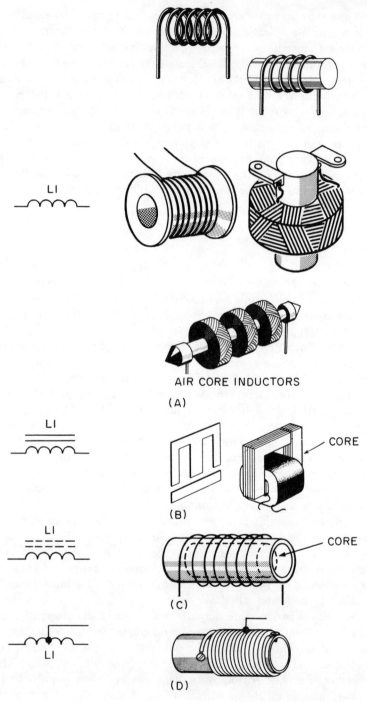

AIR CORE INDUCTORS

(A)

CORE

(B)

CORE

(C)

(D)

Fig. 2-10   *Common forms and schematic symbols for inductors.*

An inductor can be made to have more inductance with fewer turns by use of a core of permeable material such as iron or special ferrite materials. Cores are of many types. Examples are shown in Fig. 2-10(B) and (C). For low-frequency choke coils, such as are used in power supply filters, iron laminations are built up to a form suitable for mounting a layer-wound coil of many turns as shown. For coils that must operate at radio frequencies, a cylindrical form with a cylindrical ferrite core inside it is popular. In all cases of fixed cores, the symbols are as shown in Fig. 2-10(B) and (C).

Many times it is necessary that a coil be tapped. As indicated in Fig. 2-10(D), a wire lead is soldered or otherwise connected at the desired point on one of the turns. A coil may have any number of taps, as called for by the circuit. Any of the types of coils we have discussed can be tapped.

Inductors can also be made variable. The most common way to do this is shown in Fig. 2-11. A cylindrical core is inserted inside the form on which the coil is wound, and some mechanical method of moving the core in and out of the form is used. The usual way of doing this is to use a core mounted on a threaded rod, which is screwed into a threaded hole in a cap at the end of the form. Another common method is to tap the inside of the coil form to take the core, which is threaded to match. This type of core has a notch or slot in at least one end so that an adjusting tool can be used to rotate it for adjustment. These types of variable inductors are usually used for trimming or alignment adjusting purposes. Special mechanical systems have been devised for some equipment to move several cores simultaneously in and out of coils in response to rotation of a tuning dial. This *permeability tuning* technique thus substitutes for a single or ganged tuning capacitor.

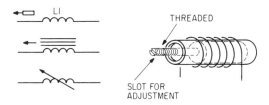

Fig. 2-11   Variable inductor and symbols.

Transformers

When an inductor is placed close enough to another inductor, the combination of the two inductors becomes a *transformer*. One inductor to which a varying or alternating voltage is applied (resulting in current through it) is called the *primary winding*. Variations in current generate a varying magnetic field whose flux lines cut across the other inductor, which is known as the *secondary winding*. Transformers are produced in a number of types and forms. In general, transformers which must operate with low-frequency currents have

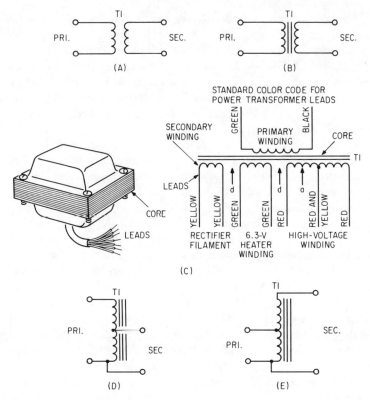

Fig. 2-12 *Transformer symbols: (A) air core, (B) laminated iron core, (C) multiple-winding power type, (D) stepdown autotransformer, and (E) step-up autotransformer.*

laminated iron cores, and those for high frequencies (RF) have either air or ferrite cores.

Schematic symbols for the most common types of transformers are illustrated in Fig. 2-12. Note the letter symbol T. In Fig. 2-12(A) is a simple air-core transformer symbol having a single secondary winding. This transformer is indicated simply by two windings alongside each other. The use of an iron core with a transformer is indicated in the same way as for an inductor, that is, with parallel lines adjacent to, and in this case, between the windings [Fig. 2-12(B)]. This indicates the laminated-iron type core, such as was illustrated in Fig. 2-10(B).

Other transformer applications are covered as they are encountered in discussions of circuits.

Switches and Relays

A means must be provided to turn equipment on and off or to alter its mode of operation. This is the job of switches, which can be either directly and

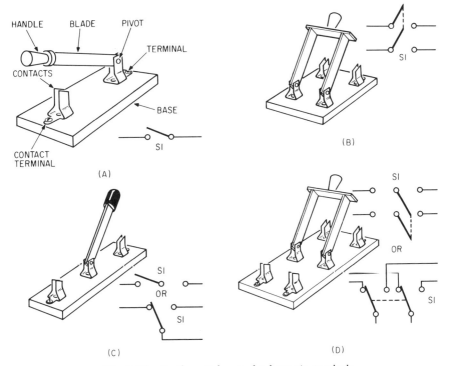

Fig. 2-13 *Knife switches and schematic symbols.*

manually operated or remotely controlled from a manual switch on the other end of an electric circuit. In the latter case, the remotely controlled switches are called *relays.*

The simplest form of a switch is a broken wire in a circuit. By placing the two broken ends of the wire together, we can start up the current (turn the equipment on) or by pulling them apart, break the circuit and interrupt the current (thus turning the equipment off). Such an arrangement would obviously be very crude; besides, when high voltages and currents are involved there could be danger from shock and heat. Also, switching must often be done in much more complicated situations than just "making and breaking" one conductor.

Perhaps the simplest of practical switches is the *knife* switch, illustrated in four forms in Fig. 2-13, along with their schematic symbols and showing the letter symbol S. The switch illustrated in Fig. 2-13(A) is a single-pole single-throw (spst) switch, and the one in Fig. 2-13(B) is a double-pole double-throw (dpdt) switch. The number of poles is the number of circuits controlled by the switch and the number of throws is the number of choices of contacts that can be made by the moving arm. Additional obvious variations whose schematic symbols are shown in Fig. 2-13 are the single-pole double-throw (spdt) in (C) and double-pole double-throw (dpdt) in (D). For the knife switch, two throws are all that are possible, but more than two poles are practical.

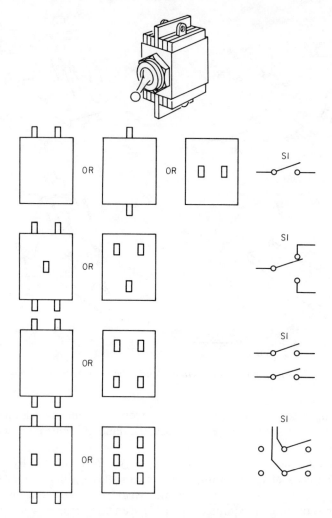

Fig. 2-14   *Toggle switch and some circuit variations.*

The knife switch has been used to introduce the schematic symbol for a switch because it is easy to correlate the physical component with its symbol. However, in modern equipment, other types are more practical. Another type, known as a *toggle* switch, is shown in Fig. 2-14. Such a switch is convenient for mounting on a panel and comes in any of the four variations, having the same schematic symbols, as for the knife switch. The figure also shows how connecting terminals may be arranged for various switch types.

When a number of throws are needed, that is, when one or more circuit leads must be connected to any of three or more other leads, a *rotary* switch is useful. Such a switch is illustrated in Fig. 2-15. The upper illustration shows a *single-deck* switch and its symbol; for more circuits, more decks can be added. A single deck can usually handle up to about 12 contacts. These can be all

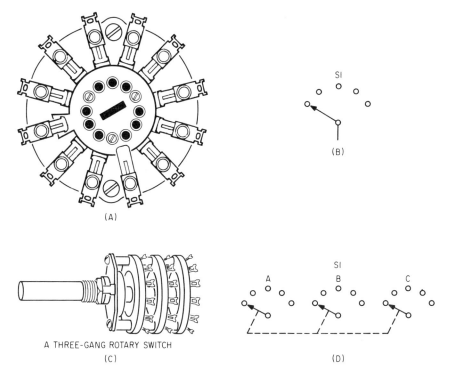

Fig. 2-15 *Rotary switches and schematic symbols: (A) and (B) single-deck type, and (C) and (D) multiple-deck type.*

selectable by a single moving contact, or divided into two or more groups, each of whose contacts can be selected by a separate moving contact. If the number of moving contacts or the number of contacts to be selected are too many for one deck, then others are added as shown in Fig. 2-15(C) and (D).

Another type is known as the *slide* switch. It is illustrated with two typical schematic symbols in Fig. 2-16. Electrically it functions exactly like the toggle and knife switches, but mechanically it operates by a sliding of the handle from side to side (or up and down, depending on how it is mounted). In the

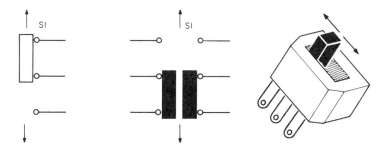

Fig. 2-16 *Slide switch and symbols for two variations.*

schematic symbol, the thin rectangular-shaped bar (or bars) moves to one side or the other, so that it short-circuits the middle contact(s) to the contact(s) on one side or the other.

As mentioned earlier, a relay is a remote-controlled switch. There are two general types of relays: electromagnetic and electronic. The letter symbol for either is K. The electronic type is more of a circuit than a component (although it is often packaged as a separate component); therefore, it is made up of transistor(s) and other components we shall be discussing later. Thus, we are interested more here in the electromagnetic type. Such relays are illustrated in Fig. 2-17. In each case, the motivating force comes from an electromagnet, operated by current switched into its windings from some location either nearby or as far as miles away. As the electromagnet pulls the armature toward it, the armature moves contacts either together or apart. The contacts may be mounted as part of the armature assembly or may be separate spring contacts operated by pressure from the armature assembly.

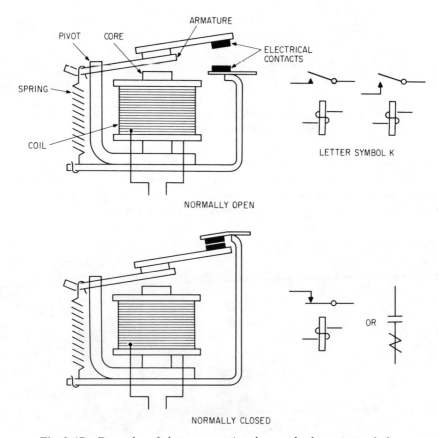

Fig. 2-17   *Examples of electromagnetic relays and schematic symbols.*

It is important to note the different arrangements of contacts that may be encountered. The most common are illustrated in Fig. 2-18. There are two basic arrangements: 1) normally open, in which there is no connection between the contacts when the relay is not operated (no current in the coil), and 2) normally closed, in which there is a connection between the contacts when the relay is not operating and this connection is broken when the current is fed to the coil and the armature operates. As can be seen from Fig. 2-18, these types can be used together on the same relay. In some cases as many as a dozen combinations of contacts are built into a single relay. Thus, from one switch, which controls the current in the coil, the functions of many switches can be controlled.

As can be seen, there are at least three advantages of relays: 1) they can combine separate functions into one, 2) they can be operated from a different chosen location, and 3) they can allow isolation of the switched circuit, avoiding such problems as hazard from high voltage and loading effects in high-frequency circuits.

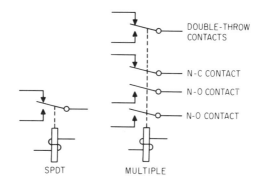

Fig. 2-18   *Types of relay contact arrangements.*

The Diode

Since vacuum tubes have almost completely been replaced by solid-state devices, the following discussions are limited to the latter. Information is included on vacuum tubes in Chap. 7, however, in connection with RF power amplifiers, where they are still frequently used.

The simplest of all solid-state components is the *diode*. The modern diode is constituted of two pieces of differently processed semiconductor material. The dissimilarity arises from the fact that, in the crystalline structure, one material contains an excess of electrons (N material) and the other has potential accommodation for electrons but has some missing (P material). When the two materials are in contact with each other to form a junction (see Fig. 2-19), they are capable of conducting electric current much better in one

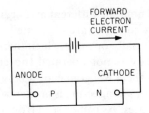

Fig. 2-19   How the junction of N- and P-type materials form a diode.

direction (forward) than in the other (reverse). Forward electron current is that entering at the cathode and leaving at the anode. Although the current is easier in the forward direction, diodes are also often used with applied voltage of reverse-current polarity and some current does result. Such applications are considered later.

The symbol for a semiconductor diode is shown in Fig. 2-20. (Sometimes the circle is not used.) The letter symbol is *D*. Very important in the interpretation of schematics with diodes is relating the direction of easy electron flow with the symbol. This is shown in Fig. 2-20. Notice that easy electron flow is in a direction opposing the arrow symbol. This is because the symbol was developed in accord with the "conventional" (positive charge) current assumption. Neither convention is "right" or "wrong" and both are widely used, and give

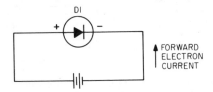

Fig. 2-20   Symbol for semiconductor diode and how it indicates direction of forward current.

the same correct answers if consistently employed. In this book, we use the electron (negative charge) flow convention and will thus see the direction of easy current as that which opposes the arrow. The relation of current to applied voltage of a typical diode is shown in Fig. 2-21.

Diodes are used as detectors, demodulators, mixers, comparators, rectifiers, and even amplifiers. One variation for use as a rectifier is the silicon-controlled rectifier (SCR) whose schematic symbol and letter symbol are shown in Fig. 2-22. The extra element is a control electrode, by which rectifier current may be cut off during parts of the applied current cycle. This device is also often used as a switch for turning circuits on by the application of voltage to the control electrode. It is discussed in greater detail later in this chapter.

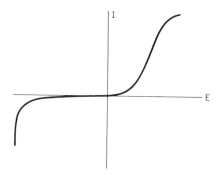

Fig. 2-21 *Graphic relationship between current and voltage for a semi-conductor diode (Courtesy, RCA).*

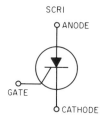

Fig. 2-22 *Silicon-controlled rectifier and its schematic symbol.*

The Bipolar Transistor

Probably the most widely used active device is the transistor. It is available (and used) in any of several different forms. The simplest is the *bipolar* transistor, which consists of three layers of specially prepared germanium or silicon (most today are silicon).

The layers of the transistor are of the same type of doped material (P and N) as used for diodes, except that the transistor has three layers. As illustrated in Fig. 2-23, the two outer layers are the same (either P or N) and the middle layer (which is relatively very thin) is of the other type. Thus, instead of just one PN junction as in a diode, there are two.

As can be further seen from Fig. 2-23, bipolar transistors come in two versions, depending upon which layer type is outside. When N type is outside, the device is an *NPN* transistor; when P type is outside, it is a *PNP* transistor.

Also important are the names of the three electrodes of the bipolar transistor. As indicated in the figure, these are the emitter (E), base (B), and collector (C), as shown in Fig. 2-23(A). The letter symbol for all transistors is *Q*.

We are interested here mainly in how transistors are shown in schematic diagrams. This is indicated in Fig. 2-23(B). Notice that it is the direction the arrow (emitter) points that determines whether the symbol indicates an NPN or PNP transistor.

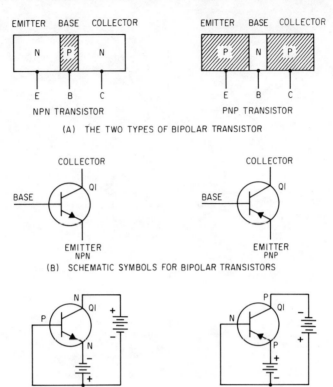

(A) THE TWO TYPES OF BIPOLAR TRANSISTOR

(B) SCHEMATIC SYMBOLS FOR BIPOLAR TRANSISTORS

(C) POLARITY OF BIPOLAR TRANSISTOR BIAS IN A CIRCUIT

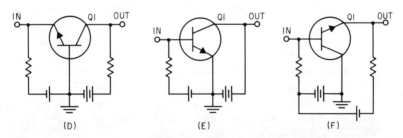

Fig. 2-23 *The two types of bipolar transistors: (A) construction, (B) schematic symbols, and (C) bias polarity. Basic circuit connections: (D) common base, (E) common emitter, and (F) common collector (Courtesy,* RCA).

In examining a schematic diagram, it is usually of interest to know voltage and current polarities and how they relate to the requirements of such components as transistors. Bipolar transistors are normally operated with the collector *reverse-biased* and the emitter *forward-biased* with respect to the base. P stands for positive, so that P material is forward-biased when it is more positive than the base. Likewise, N material is forward-biased when it is negative with respect to the base. The opposites are reverse-biased. This relationship is true no matter which of the three elements is grounded; the relative relationship is the same.

Consider the drawing, Fig. 2-23(C). This shows the schematic symbols of the two types of bipolar transistors and the bias polarities with which they are normally operated. At the left, the NPN transistor needs a positive voltage on the collector and a negative voltage on the emitter. At the right, with the PNP transistor, the conditions of polarity are reversed.

Also important is how a transistor is functionally connected into a circuit. There are three possible basic arrangements.

1. **Common base.** From the standpoint of structure, this one seems most natural. The base is the common circuit reference and is usually operated at ground potential as far as signal is concerned. A basic schematic diagram is shown in Fig. 2-23(D). The output of this circuit is in phase with its input, and gain is relatively low.

2. **Common emitter.** This is the most often encountered arrangement, being the analog of the common-cathode tube circuit. The emitter is normally at signal ground but often has a dc bias Fig. 2-23(E). The voltage output is 180° out of phase with the voltage input and the gain is the highest of the three basic arrangements.

3. **Common collector.** This is the analog of the tube cathode follower and is good for matching high-impedance inputs to low-impedance outputs [Fig. 2-23(F)]. Its voltage gain is lowest, but it can provide high power gain.

Field Effect Transistor

Although bipolar transistors are the most common, another type—the *field effect transistor*—is also widely used. It overcomes what is often a disadvantage in bipolar transistors: low-impedance inputs and outputs. The input current path is continuous through the bipolar transistor, and this represents a low input impedance. In the field-effect transistor (FET), however, the output current is controlled by the electric field of the input signal. This field induces a charge in a semiconductor layer (channel) which is part of the output circuit. The induced charge controls the output current. Since it is the *voltage* applied to the input electrode (gate)—and not the current into it—that controls the output current, high input impedances are achieved.

There are two general types of FETs: the junction-gate type (JFET) and the metal-oxide semiconductor type (MOSFET). The construction and schematic symbols for the two variations of the JFET are shown in Fig. 2-24. These variations are distinguished as N type or P type, depending upon which type of semiconductor material is used in the channel. In Fig. 2-24(A) is a cross-

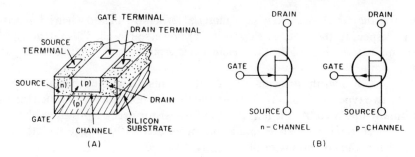

Fig. 2-24   *Construction drawing and schematic symbols for two variations of JFETs (Courtesy,* RCA).

sectional drawing of the construction of the N type. The gate, which in this type is of P type material, is embedded in N type material. The portion of the N type material under the gate is very thin and is called the *channel*. The two edge portions of the N type material are called the *source* (left) and the *drain* (right). Output current appears between the source and the drain through the channel. This current is controlled by the voltage on the gate. The gate and channel form a PN junction. The junction is reverse-biased and the level or "amount" of this bias controls the source-to-drain output current. A P channel JFET works the same way, except that the Ps and Ns are interchanged and the biasing and output current are reversed. The schematic symbols for the two JFETs are shown at the right in Fig. 2-24.

The other general type of FET is the *metal-oxide semiconductor* (MOS) type. The MOSFET can be of either the *enhancement* or *depletion* type. Each of these can be of either the N type or P type, depending upon the material used in the channel. The construction of an N-type MOSFET of the enhancement variety

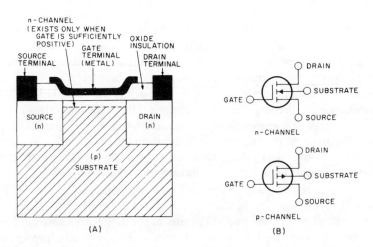

Fig. 2-25   *Construction drawing and schematic symbols for an enhancement-type MOSFET (Courtesy,* RCA).

is shown in Fig. 2-25(A). It can be a P type of the same variety if the N material and P material are swapped. The schematic symbols for both N and P types are shown in Fig. 2-25(B). Note in Fig. 2-25(A) that a thin layer of oxide separates the metal gate from the semiconductor material. The oxide material is a very good insulator and allows a very high input resistance for this type of transistor. When there is sufficient bias on the gate, an N-type region is formed in the top of the N-type material, causing source-to-drain conduction. In this type of transistor, there is no conduction unless there is positive bias voltage on the gate. Notice that in the schematic symbols for this type, the "channel line" at the right is a broken line, to symbolize the lack of conductivity for zero or reverse gate bias.

The construction drawing and schematic symbols for the depletion type of MOSFET are shown in Fig. 2-26. The difference here is that the N channel is permanent and exists even when forward bias is not applied. There is normally some conduction even at zero bias and appreciable reverse bias is necessary to achieve cutoff. This is symbolized in the schematic symbols in Fig. 2-26(B) by the continuous channel line.

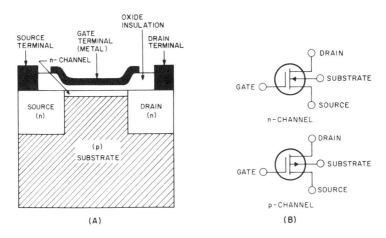

Fig. 2-26 *Construction drawing and schematic symbols for a depletion-type MOSFET (Courtesy, RCA).*

Depletion-type MOSFETs with two gates are also used. The construction drawing and schematic symbol are shown in Fig. 2-27. The two isolated gates make this type very useful where signals from two separate sources are to be combined, such as in gain-controlled amplifiers, mixers, and demodulators.

Other Semiconductor Components

Although diodes and transistors make up most of the semiconductor devices in use, there are a number of others that we should also consider.

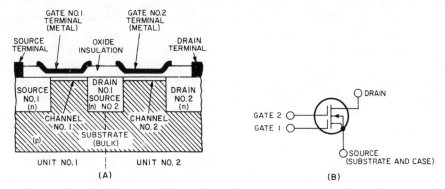

Fig. 2-27   *Construction drawing and schematic symbols for a dual gate depletion-type MOSFET (Courtesy, RCA).*

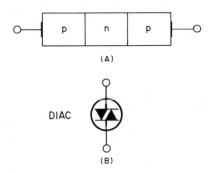

Fig. 2-28   *Construction drawing and schematic of a diac (Courtesy, RCA).*

One is the *diac*. The construction drawing and schematic symbol for this device are shown in Fig. 2-28. It is analogous to two diodes connected in series, except that the N portion is common, forming a PNP arrangement with two PN junctions. Unlike the case of a transistor, the junctions in this case have the same electrical characteristics so that the device is electrically symmetrical. There is no widely accepted or used letter symbol for the diac; it is usually identified on the schematic with its manufacturer's type number.

The diac can operate with voltages of either polarity. A rising voltage of either polarity meets a relatively high resistance until the reverse-biased junction undergoes what is called *avalanche breakdown*; then its resistance goes to a low value and current increases with decreasing voltage for a negative resistance characteristic. This is illustrated by the device's current–voltage curve, shown in Fig. 2-29. Diacs are useful for triggering thyristors and other devices in such applications as motor control and lamp dimming.

Other devices fall under the heading of *thyristors*. These are devices using several alternating layers of P-type and N-type semiconductor materials. Types of thyristors include *silicon-controlled rectifiers* (SCR), *triacs*, and *bilateral switches*.

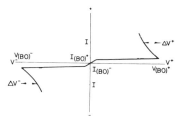

Fig. 2-29   *Current-voltage curve for a diac* (*Courtesy*, RCA).

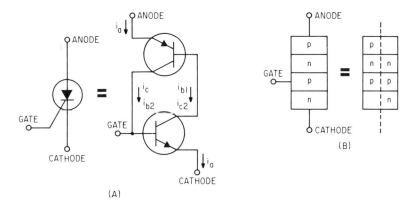

Fig. 2-30   *Schematic symbol of an SCR and equivalent structural arrangement* (*Courtesy*, RCA).

The SCR is constructed essentially of four alternating layers of P- and N-type semiconductor materials. It is somewhat like a transistor with the triggering element (gate) added. The schematic symbol and essential construction are shown in Fig. 2-30. Notice that the three terminals are labeled *anode*, *cathode*, and *gate*. The characteristic function of an SCR is as follows:

1. A forward (anode positive) voltage is applied between the anode and cathode.
2. A positive voltage is applied to the gate. If this voltage is equal to, or greater than, a given *trigger level* it injects current into the device to switch the main anode-to-cathode circuit into a conducting state.
3. Once the triggering takes place, the gate loses control and the anode–cathode current continues until the anode–cathode voltage falls to near zero or reverses its polarity. As indicated in Fig. 2-30, the action is the same as that of two transistors connected as shown. A typical voltage-current characteristic is shown in Fig. 2-31.

Typical SCR applications include motor controls, power supply regulation and phase control, choppers and inverters, and static switching in

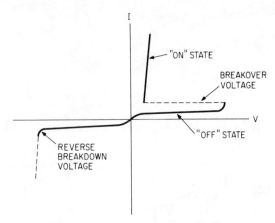

Fig. 2-31   *I–V characteristics of an SCR.*

instrumentation. As previously stated, the letter symbol for an SCR is the same as its abbreviation: *SCR*.

The *triac* configuration is shown in Fig. 2-32, which also illustrates the schematic symbol for this device. Notice that the symbol is the same as for the diac except that the triac has a gate terminal. The triggering is similar to that for the SCR. However, unlike the SCR, the triac current cannot be cut off by polarity reversal of the main anode–cathode voltage. This device is bidirectional, that is, it will operate with the main voltage of either polarity. Therefore, it is like

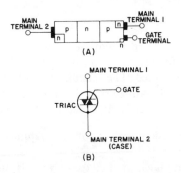

Fig. 2-32   *Schematic, symbols, and construction of a triac* (*Courtesy, RCA*).

an SCR in either direction. Only when the main voltage (of either polarity) is reduced below the minimum level required to hold conduction can the main current be cut off. A typical characteristic is shown in Fig. 2-33. Triacs are used in a wide variety of ac power control applications.

The *bilateral switch* is made up of four layers in a PNPN configuration and has two gates: an anode gate and a cathode gate. The configuration and the

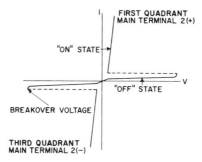

Fig. 2-33   *Typical characteristics of a triac (Courtesy, RCA).*

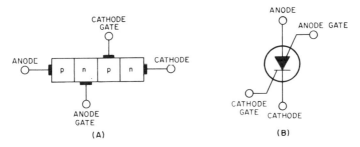

Fig. 2-34   *Configuration and schematic symbol of a bilateral switch (Courtesy, RCA).*

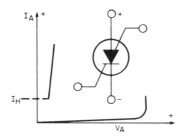

Fig. 2-35   *Characteristics for a bilateral switch (Courtesy, RCA).*

schematic symbol are illustrated in Fig. 2-34. This device can be turned on by increasing the anode–cathode voltage to its breakdown value, or keeping it at a lower voltage and feeding current into either of the gates. A typical characteristic is shown in Fig. 2-35. Bilateral switches are used in such applications as binary counters, shift registers, relay drivers, and indicator lamp drivers.

As in the case of the diac, triacs and bilateral switches have no widely accepted or used letter symbol and are usually identified on the schematic diagram by the manufacturer's type number.

Simplest Discrete Component Circuits

To this point, we have been considering schematic symbols for individual components. Before proceeding to complete schematic diagrams, it is desirable that we first consider significant sections or units into which schematic diagrams can be divided. These divisions are sometimes called *building blocks*. They each perform a particular function. A given building block may be used only once in a circuit; others may appear two, three, or any number of times.

The following discussion covers a number of the more common circuit building blocks, to show how the representation of components is extended to an arrangement containing several circuits.

It must be noted that these building blocks are not necessarily physically separate entities. They are functional entities, and are often integrated into the overall circuit. But if we are to analyze a schematic diagram, we must be able to break it down into its significant parts—its building blocks. These should not be confused with physical entities, such as modules and integrated circuits. These are considered later in this chapter.

Probably the simplest building block is an arrangement of resistors. As indicated in Fig. 2-36, two or more resistors may be connected in series, parallel, or series–parallel. Such an arrangement may be a functional entity on its own, or can be a part of a more complex building block.

Typical simple resistor arrangements and ways of showing them schematically are illustrated in Fig. 2-37. In Fig. 2-37(A) is shown three variations of the schematic diagram of two resistors in series. Notice that,

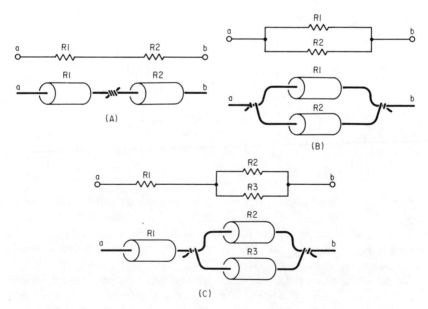

Fig. 2-36   *Resistors connected in series, parallel, and series–parallel.*

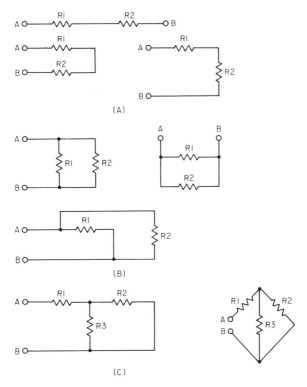

Fig. 2-37 *Illustrations of how different schematic symbol layouts can be used for the same circuit: (A) series, (B) parallel, and (C) series-parallel.*

although the physical arrangement of the symbols is a little different in each case, an electrical trace from A to B would pass through the resistors in the same sequence and the total current passes through both. The small circles marked A and B stand for terminals, or points at which other components or circuits are (or may be) connected. If these circuits were included as part of a larger one, the terminals would normally not be there. In some diagrams, of course, these symbols represent actual physical terminals.

In Fig. 2-37(B) are ways of showing two resistors in parallel. In each case an electric current, entering at (A) and leaving at (B), would divide between the two resistors. This action is the same in all of the arrangements, even though the physical orientation of the symbols differs. In the same way, at Fig. 2-37(C) are two variations of a series-parallel resistor circuit diagram. The two variations are electrically identical.

Not only resistors can be connected these ways, but also capacitors or inductors, or certain other components. The principles are exactly the same as indicated above for resistors.

Resonant Circuits

When an inductor and a capacitor are connected in series or parallel, the combination has a property called resonance. That is, the impedance of the circuit is sensitive to the frequency of the ac voltage applied to it. Since this property is fundamental and is widely applied in electronic equipment, the resonant circuit can be considered a basic building block.

The two basic resonant circuit arrangements are shown in Fig. 2-38. In the series resonant circuit, Fig. 2-38(A), note that the current from A to B must be carried by both L and C. The frequency of resonance for L and C is

$$f = \frac{1}{2\pi\sqrt{LC}}$$

where f is in hertz, L is in henries, and C is in farads.

This is the same frequency at which the reactance of L is exactly the

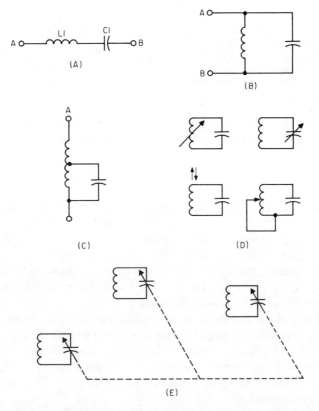

Fig. 2-38 *Two basic resonant-circuit configurations: (A) series and (B) parallel. Also, (C) tapped resonant circuit, (D) common tuning arrangements, and (E) ganged tuned circuits.*

same as that of C. In the series resonant circuit, the reactance of L and C cancel and the total impedance (A to B) is near (theoretically equal to) zero. In Fig. 2-38(B) is a parallel resonant circuit. Here the effect is the opposite, with the impedance between a and b building up to a relatively high value (theoretically infinite) at the resonance frequency. Sometimes the capacitor is connected across only part of the inductor, as shown in Fig. 2-38(C).

The importance of the resonant circuit lies in its use for tuning circuits in radio and TV receivers and transmitters, test equipment, and a wide variety of other electronic applications. For tuning, the frequency at which resonance occurs must be made variable and adjustable. It may be either the capacitor or the inductor (or both) that is made variable. Tuning circuits are usually shown in a schematic diagram as illustrated in Fig. 2-38(D). The most common arrangement is that in which the capacitor is variable (upper right). In other instances, the inductor is made variable and the capacitor is fixed. (The ways of making capacitors and inductors variable was discussed earlier in this chapter.)

Often it is necessary, as mentioned earlier, to *gang* different tuned circuits so that they will all work together. With variable capacitance, this can be done with a ganged capacitor (described earlier). Ganged tuned circuits are shown on a schematic diagram by connecting the variable components (in this case the capacitors) together with a dashed line, as illustrated in Fig. 2-38(E). When the inductors are variable, the method is the same, except that the arrows showing inductor variability are connected with the dash lines.

Amplifiers

Probably the most common of electronic building blocks is the amplifier. In contrast to the circuit sections discussed so far, which are called *passive* circuits, the amplifier is called an *active* circuit, as are all circuits in which power is added so that signal current, voltage, or power output is increased in amplitude. An extremely large number of types of amplifiers is in use. No attempt will be made here to cover each type completely. Rather, examples of the more common types will aid in their recognition on schematic diagrams. Other, more special types, are discussed as they are encountered in later chapters.

The simplest type of amplifier is a resistance-coupled type using a bipolar transistor. The schematic diagram of such an amplifier is shown in Fig. 2-39(A). Notice that it uses a PNP transistor. The collector is reverse-biased because a negative voltage is applied through load resistor R_L. The negative voltage applied through the base resistor to the base makes the grounded emitter positive with respect to the base and is thus properly forward-biased. Either of the other transistor orientations (common-base, common-collector) explained earlier in this chapter could apply with the same type of circuit, but the common-emitter arrangement shown here is the most often encountered. If an NPN transistor is used, the voltages are reversed.

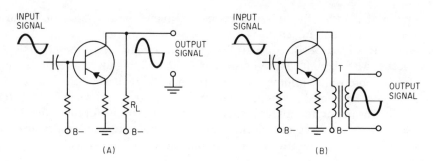

Fig. 2-39   *Basic amplifier circuits.*

Transformer-coupled amplifiers also are widely used. A typical amplifier of this type is diagrammed in Fig. 2-39(B). The collector current passes through the primary winding of a coupling transformer. The ac signal portion of the current through the primary induces voltage in the secondary winding. This voltage may then be applied to the base of another transistor for additional amplification.

Either of these amplifier types may be used in cascade, that is, with the output of one feeding the input of the next to increase amplification. The amplification of the combination is the product of the amplifications of each stage. For example, two cascaded stages each having an amplification of 10, together have an amplification of 100 ($10 \times 10 = 100$).

A simplified diagram of a typical two-stage resistance-coupled amplifier is shown in Fig. 2-40(A). Notice that each stage is the same as the individual amplifier in Fig. 2-39(A). The two stages, as they are called, are cascaded by coupling the signal being amplified from the collector of the first stage through a capacitor to the base of the second stage. A capacitor is used so that the collector-base dc voltage relationship will not be disturbed.

One of the most common applications of transformer coupling and cascaded stages is the *intermediate-frequency (IF)* amplifier. A typical simple IF amplifier is illustrated in Fig. 2-40(B). As we shall see in Chapter 6, the function of the IF amplifier is to take the signal from the mixer, amplify it, and deliver the amplified signal to the detector. Notice that in this case, instead of a capacitor, a transformer is used to transfer signal from one stage to the next. The use of two capacitors across the secondary winding allows the connection from the base of the next transistor to be "tapped down" on the secondary. The capacitors form a *voltage divider* so that the low-impedance transistor base circuit does not load down the transformer; thus, the voltage divider improves the impedance match. The arrows through the windings of the transformers show that these inductances are variable for tuning purposes. The capacitors connected in the emitter circuits and between B− and the transformer primaries are decoupling capacitors. They prevent signal voltages from building up across the resistors in the power supply circuits. Such buildup causes degeneration which reduces the gains of the stages.

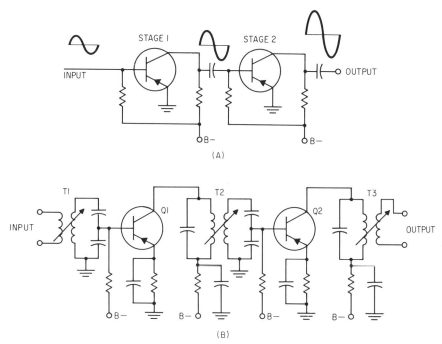

Fig. 2-40 *Cascaded amplifier: (A) resistance-coupled, and (B) transformer-coupled.*

It is not unusual to find not only bipolar transistors but also JFETs and MOSFETs used as amplifiers. Examples of a JFET type and a MOSFET transistor in amplifier circuits are shown in Fig. 2-41(A) and (B), respectively. Notice, that, in this type of circuit, which is the same in principle as the common-emitter arrangement of Fig. 2-40, the source takes the place of the emitter and the drain of the collector. Thus, these are often referred to as *common-source* circuits. Similarly, there are common-base and common-drain circuits, but the common-source is by far the most common.

It is common for audio output power amplifiers (and sometimes also preceding driver stages) to use a *push–pull* amplifier circuit. This is desirable, not only so that two transistors share the load, but also because the balanced configuration causes even-harmonic distortion to be cancelled out, thus greatly lowering overall distortion.

In the push–pull amplifier, the two transistors must operate on signals 180° out of phase with each other. If the previous stages are *single-ended* (that is, one transistor), there must be some way the phase can be changed by 180° (inverted) for one of the push-pull transistors. A device to accomplish this is called a *phase inverter*. If transformer coupling is used between the previous stage and the push-pull amplifier, the transformer is usually the phase inverter. An example of a transformer-coupled push-pull amplifier is shown in Fig. 2-42(A). The secondary winding of the input transformer is center-tapped. The signal

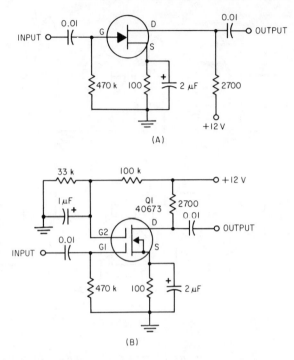

Fig. 2-41   *Amplifier circuits using FETs.*

voltages at the two ends of this winding are 180° out of phase, and are thus appropriate for the base inputs of the two transistors to which they are connected. R2 and R1 form a voltage divider across the 12-V supply; thus, R1 is used to adjust the base voltage for proper collector current. A conventional push-pull output transformer is used.

Other phase inversion methods include use of a separate transistor to shift the phase or the pickup of the additional signal from the emitter circuit. Perhaps the most common is the complementary-symmetry method (Fig. 2-42(B)) in which the differences between polarity relationships of an NPN and a PNP transistor are used.

Oscillators

There is one other basic building block circuit—the *oscillator*. Like amplifiers, oscillators are used in many types of equipment for a wide variety of functions. As with amplifiers, there is a wide variety of oscillator circuit types. Here we'll present a few of the most common circuits, emphasizing those aspects most useful in the identification of each type in schematic diagrams.

In most cases, an oscillator is just an amplifier with the output fed back to the input in such a way as to reinforce that input (positive feedback). For oscillation the feedback must be sufficient to overcome losses in the circuit.

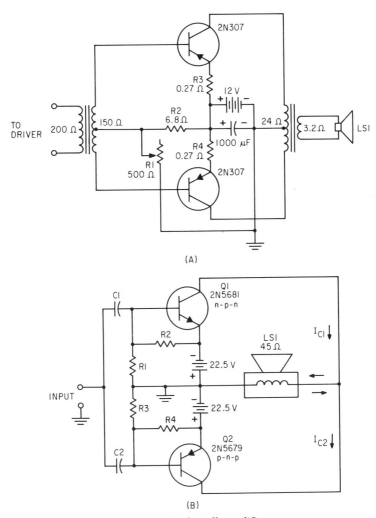

(A)

(B)

Fig. 2-42   *Push-pull amplifiers.*

An oscillator circuit can be identified primarily by the feedback arrangement. Different oscillator circuits have been given names according to the scheme used for accomplishing the feedback or the name of the person who developed the circuit. One of the best known is the *Hartley,* illustrated schematically in Fig. 2-43(A). The feedback is provided by coupling in coil L. The function is clarified by the simplified circuit in Fig. 2-43(B). This diagram shows only the RF circuit, ignoring the direct current paths. Notice that the collector signal current passes through the lower portion of L. This signal is coupled into the remainder of the coil and is thus fed back to the base.

Another oscillator circuit, which in a way is just a variation of the Hartley, is illustrated in Fig. 2-44(A). The principle is the same but here the

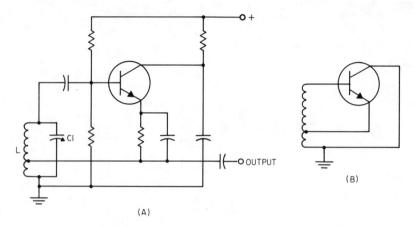

Fig. 2-43   *Hartley oscillator: (A) circuit, and (B) RF equivalent.*

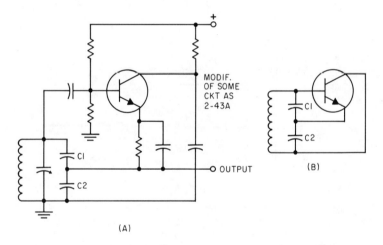

Fig. 2-44   *Colpitts oscillator: (A) circuit, and (B) RF equivalent.*

emitter circuit, instead of being tapped directly to the coil, is "tapped" by the voltage divider formed by capacitors C1 and C2. This is called the *Colpitts* circuit. The diagram in Fig. 2-44(B) shows the simplified RF circuit.

Another feedback arrangement uses an RF transformer with two separate but coupled coils, as shown in Fig. 2-45(A). Collector circuit current in L1 induces a feedback voltage in L2, which is connected to the base. This is sometimes called the *Armstrong* oscillator and L1 is often referred to as a *tickler* coil. The simplified RF version is shown in Fig. 2-45(B).

The above oscillators use tuned circuits and are commonly used for generating signals at radio frequencies. For the audio frequency range, resistance coupling is often used. Examples are shown in Fig. 2-46. The circuit of

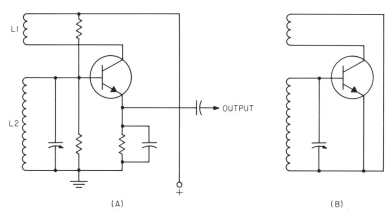

Fig. 2-45   *Armstrong oscillator: (A) circuit, and (B) simplified RF circuit.*

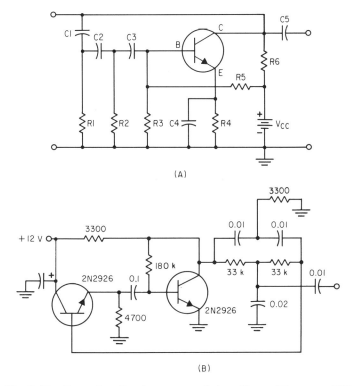

Fig. 2-46   *Examples of resistance-coupled oscillators (Courtesy, RCA).*

Fig. 2-46(A) is called the *phase-shift* oscillator. The collector signal is fed back through C1 and R1. C1-R1, C2-R2, and C3-R3 form a phase-shifting network which shifts the fed-back collector signal phase 180° so that it arrives at the base in phase with the signal at the base. The oscillation frequency is adjusted by changing the values of resistance and capacitance in the phasing network to give 180° shift at the desired frequency.

Another resistance-coupled oscillator, in this case using two transistors, is shown in Fig. 2-46(B). Signal from the output of the second transistor (a common-emitter amplifier) is fed back to the base of the first transistor (a common-collector amplifier). The output goes to the base of the second transistor, completing the feedback path. In the output collector circuit is a *twin-T* phase-shift circuit that controls frequency.

Another common class of oscillator circuits includes those known as *relaxation* oscillators. An example is shown in Fig. 2-47. It is called a *blocking* oscillator. The transformer couples positive feedback. As feedback builds, the transistor saturates, causing the feedback to fall off. Capacitor C discharges from base to emitter, then starts to charge again. The result is a series of pulses, as indicated. The widest use of this type of oscillator is in oscilloscopes and television receivers, where it is used as a vertical or horizontal sweep oscillator producing a sawtooth-waveform signal. More examples are given in Chap. 8.

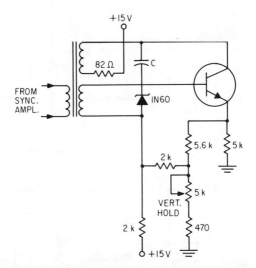

Fig. 2-47   Blocking oscillator.

Linear Modules

Thus far we have discussed components and the electrically significant groupings of those components (resonant circuits, amplifiers, etc.). Another important grouping is the *module*. A module is significant not only electrically but also physically in that it can easily be separated from an overall assembly and replaced without any (or with only a minimum) of wiring work. Modular construction has now become virtually universal in television receivers and in computers. Typical assemblies would be a power supply, an IF amplifier section, or a vertical deflection section. The circuits employed in modules are of the same type as those used in nonmodular equipment, examples of which have already been discussed.

Often the only indication of a module on a schematic diagram is a dashed line enclosing its contents. Sometimes modular entities are separated. Otherwise, they may not be indicated on the schematic diagram. However, an additional layout diagram will often be included to show the modular design. Modular construction is closely related to the specific equipment design in which it is included. For this reason, discussion of modules will be included in later chapters with equipment to which they apply.

Digital Logic Packages

Computers and computer-type control equipment use vast numbers of digital logic packages. These are individual circuit sections, which, either alone or in groups, perform logic functions. Each package is classified according to its function and assigned a schematic symbol. Even though a package contains many components, it is still designated by a basic logic package symbol. In general, computer diagrams contain only these logic symbols. However, when the design of a package itself (as opposed to the combination of packages) is discussed, a conventional schematic diagram of its contents is required. Of course, interconnection wiring diagrams are used in computer construction.

We shall discuss first the nature of each of the common types of logic packages, something about their circuits, and how logic packages are used to build up computer type circuits.

We refer to *digital logic*. This means that the signals handled are *digital* signals—signals that can have either of two values: a maximum voltage or current and a zero or minimum voltage or current. One state is designated a *1* and the other a *0*. Usually the full current or voltage is designated *1* and zero current or voltage is designated *0*, but the reverse is sometimes true.

The logic function depends on the relationship between the output of a package and the inputs (normally more than one) that produce that output. Each type of logic function package has its own schematic symbol and in logic type schematic diagrams it is these symbols, rather than resistors, capacitors, etc., that are shown. Because the function of this type of logic package is to be either *open* or *closed*, it is often referred to as a logic *gate*.

Logic Gate Circuits

Before proceeding to a list of the functions and logic schematic symbols, consider the basic contents of some simple logic gates. This will provide an idea of the electrical function so that handling of packages is more understandable. The evolution of a typical OR circuit from simple resistor logic to diode-transistor logic is shown in Fig. 2-48.

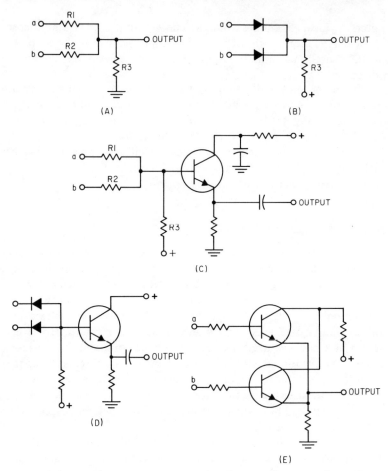

Fig. 2-48 *Examples of logic circuitry in OR gates: (A) resistor logic (RL), (B) diode logic (DL), (C) resistor-transistor logic (RTL), (D) diode-transistor logic (DTL), and (E) transistor-transistor logic (TTL).*

   In Fig. 2-48(A) is a simple resistor logic (RL) circuit. This circuit is purely for discussion; it is seldom used as shown because of its losses and lack of isolation between output and input. However, it illustrates the principle of an OR gate. The idea of an OR gate is that there is an output if there is an input on either *a* or *b* or both. Note that for input *a* there is a voltage divider, R1-R3, to ground and for input *b* a voltage divider, R2-R3. Thus, a pulse (1) on either input (or both) results in an output pulse across R3. For reasons given above, this type is hardly ever used.
   The next step is diode logic (DL), illustrated in Fig. 2-48(B). Here R1 and R2 have been replaced by diodes and a bias has been applied to R3. This bias is applied to the negative terminals of the diodes through R3 so that, in the

absence of input, the diodes are cut off and there is no output. If a positive pulse is applied to either *a* or *b*, the diode involved is turned on and conducts through R3 (the input pulse is appreciably higher than the bias on R3).

Diode logic provides some isolation between input and output, which, because of the sharp on–off characteristic of the diodes, keeps input noise problems down. However, the forward resistance of the diodes causes sufficient loss so that several gates in series cannot be used. For this reason, the most common logic gates use transistors.

Illustrated in Fig. 2-48(C) is resistor-transistor logic (RTL). Here, to the resistor logic gate of Fig. 2-48(A) is added a transistor emitter-follower. A positive input at either *a* or *b* overcomes the bias and turns on the transistor, causing the collector-emitter circuit to conduct. Output is taken from the emitter because the voltage at that point is in phase with the inputs. The advantage of this circuit over those in Fig. 2-48(A) and (B) is that the amplification of the transistor makes up for the loss in the resistors. Any reasonable number of gates thus may be used in cascade without loss of signal energy.

In Fig. 2-48(D), the diode logic arrangement of Fig. 2-48(B) is followed by a transistor. The principle is the same as in Fig. 2-48(C). It is called diode-transistor logic (DTL).

In Fig. 2-48(E), the two transistors become the basic logic elements. In essence, the two transistors are connected in parallel except for their inputs (bases). An input at either *a* or *b* causes one of the transistors to turn on and produce output. This is called transistor-transistor logic (TTL).

The diagrams of Fig. 2-48 give an idea of how logic operations are accomplished. For simplicity, we have shown in Fig. 2-48 only gates with two inputs. Such circuits are by no means limited to two, but may have any number, many having three or four or more.

There are many variations of these circuits but the principle is the same. An understanding of this principle should allow one to recognize other circuits that may be encountered. Having now considered the types of internal logic gate circuitry, we can proceed to a consideration of these gates as packages, in order to show how they are used in logic diagrams.

Types of Logic Packages

In order to consider the operation of different logic packages, it will be helpful to become familiar with *truth tables*. These tables, often found on schematic diagrams or closely used with them, tell what the output is for each combination of inputs for the applicable package. Let us first consider the truth table for the OR package illustrated by the circuits of Fig. 2-48.

Such a truth table is illustrated in Fig. 2-49. The first row of numbers shows that if inputs a and b are both 0, then the output is 0. The second row shows that if a is 1 and b is 0, then the output is 1. The third row shows that the result is the same if b is 1 and a is 0. The last row shows that two input 1s result in a 1 output.

INPUT		OUTPUT
A	B	C
0	0	0
I	0	I
0	I	I
I	I	I

Fig. 2-49   Truth table for OR package.

Other types of logic packages, truth tables, and basic schematic symbols are illustrated in Fig. 2-50. First listed is the OR package. Its truth table has already been discussed. The symbol is shown in the right-hand column.

Listed second in Fig. 2-50 is the AND package. This package gives output only if there is a 1 on both inputs at the same time. Thus, AND means that there must be inputs on a *and* b (*and* c, *and* d, etc., if there are more than two inputs) to produce a 1 output. Thus, in the truth table, there is a 1 output only when there are 1s on both (or all) inputs. Notice the particular schematic symbol for an AND gate, shown at the right.

The OR and AND packages are the most fundamental. In modifying these packages, and building up logic circuits, the inverter (INV) package is important and is listed next in Fig. 2-50. This package simply reverses input: if the input is a 0, the output is a 1; if input is 1, it puts out a 0. If the inverter is a separate package, it is drawn as at the left. If the inversion *function* is part of another package, it is shown as a small circle at the right in the figure. As we shall now see, this is used to adapt OR and AND packages to other functions.

It should be noted that a separate inverter package is not always needed, because inversion is an inherent part of a package. For example, the OR circuits in Fig. 2-48(C), (D), and (E) become NOR circuits if the output is taken from the collectors instead of the emitters.

The first example of such adaptation is the *NOR* gate, which stands for Not-OR. In logic terminology, since the only states are 0 and 1, a *not-0* is one and a *not-1* is zero. In other words, where the signal flows through a small circle, it comes out as the opposite. The term *NOT* is also applied to complete packages. In this case a NOR is a package that puts out a 1 when an OR output would be a zero and a 0 when an OR would put out a one.

Notice the truth table for the NOR circuit in Fig. 2-50. Since the first row is the only one that would give a 0 output in an OR situation, it is the only one that gives a 1 output in the NOR case. Notice that the output of the NOR truth table is the opposite of that of the OR. Observe now the first NOR schematic symbol at the right. It is really made up of an AND gate and inverters. Knowing that the unique case in the NOR truth tables exists whenever the inputs are both 0, let us test the symbol. The two 0 inputs are converted to 1s by the inverters. The AND that follows properly produces a 1 output. Since in every other case at least one input is a 1, which is inverted to 0, the two inputs to the

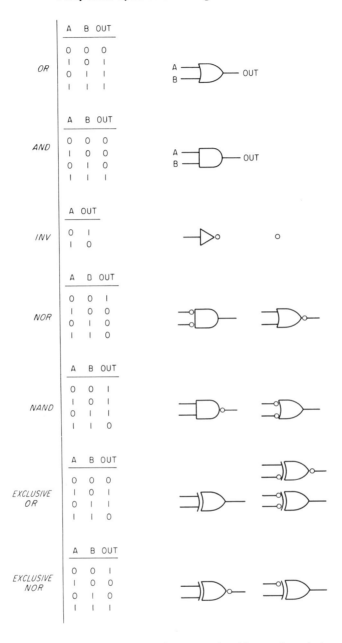

Fig. 2-50   *List of logic packages, truth tables, and symbols.*

AND are never both 1s, so all other outputs are 0. As shown by the NOR symbol on the right, an OR can be used to make a NOR by simply putting an inverter in the output.

Now consider the next listing in Fig. 2-50, the *NAND*. As might be expected, the truth table is the AND table with 1s and 0s swapped off in the outputs. Also, the symbol (left diagram) can be simply the AND symbol with an inverter in the output to "not" the AND. The right-hand diagram shows how the same function can be expressed by an OR with an inverter in each input. It can be seen from the table that this forms a NAND.

There is an important variation of the OR. We have said that the OR produces an output if there is a 1 on either input or on *both inputs*. Sometimes an OR that does not respond to both inputs is desired. This is known as an EXCLUSIVE OR, whose truth table and symbols are shown in the next listing of Fig. 2-50. Note that the symbol is that of an OR with an arc in front of it. This symbol can be used as is (left symbol) or with inverters as shown at the right.

The EXCLUSIVE NOR is the final listing in Fig. 2-50. Note the swapping off of 1s and 0s in the output as compared to the EXCLUSIVE OR. Also the simplest symbol: an EXCLUSIVE OR with an inverter in the output. As shown at the right, an EXCLUSIVE OR with an inverter in one input gives the same function.

Universal Gates

In Fig. 2-50, we show how some basic gates can be modified to form other gates. This is important because of the way gates are made available on some integrated-circuit packages (*ICs*) which are discussed presently. To make a given IC more versatile, it is often designed with a number of basic transistor circuits, so these can be applied to any of many requirements by the way the externally available terminals are connected. For example, it was shown that a NOR could be made by combining either an AND with two inverters or an OR with one inverter.

Examples of how this can be done with NANDs and NORs are shown in Fig. 2-51. Notice in the first row that even an inverter can be made from a NAND or a NOR, by adding inverters or inverter functions. Using the principles of NAND and NOR functions already discussed, the makeup of the other combinations can be verified by following through each combination functionally. Thus, if an IC or other circuit chip includes only a number of NANDs or a number of NORs, these can be connected together to form whatever other functions are desired.

Sequential Logic Packages

Up to now, we have been talking only about *combinatorial* logic. This is logic in which circuit response at any given moment is a result only of the combination of inputs then present. On the other hand, *sequential* logic is that in which circuit outputs depend not only on present inputs but also on earlier inputs. In other words, in a manner of speaking, a package "remembers" what went before in a sequence of events. Thus, the circuits we are now to consider may be referred to as *memory* type circuits.

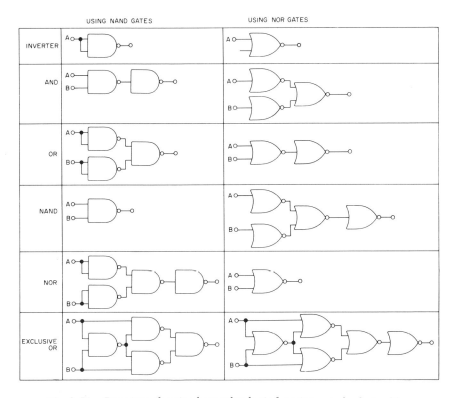

Fig. 2-51   *Drawings showing how other logic functions can be derived from NANDs and NORs.*

The most common example of this type of circuit is the *flip-flop*, a special form of multivibrator. Like other packages, it has two states, which might be described as on and off, (or 1 or 0). When it receives a pulse (1) it operates, but its output depends on what its state is when the pulse arrives. If its state is 0, it becomes a 1; if it's a 1, it becomes a 0. Unlike a multivibrator oscillator, which continues to oscillate between the two states, the flip-flop changes state once only when a new pulse arrives from an external source.

A typical flip-flop circuit is shown in Fig. 2-52. Note that it consists of two RC amplifiers, with the output (collector) of each coupled to the input (base) of the other. There are two inputs, labeled R and S, and two outputs, Q and $\overline{Q}$. (The bar over the Q is the Boolean algebra symbol for *not*.) In normal operation the $\overline{Q}$ output is the inverse of Q. In this case, the inputs R and S are shown as switches. The letters stand for *reset* and *set*, and this circuit is often referred to as an *RS* flip-flop. Thus, in the use of this circuit an input of a 1 would be a short circuit. The operation of the circuit is such that there are only two possible states: (1) Q1 conducting and Q2 cut off, and (2) Q2 conducting and Q1 cut off.

Suppose the circuit starts with condition (1). With Q2 not conducting, its collector voltage is high (near Vcc) and output Q is 1. Since Q1 is conducting,

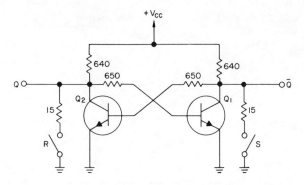

Fig. 2-52   *Flip-flop circuit.*

its collector voltage is near zero, so that output $\overline{Q}$ is 0. Now suppose switch R is closed. This drops the collector voltage of Q2 and base voltage of Q1 to near zero, causing Q1 to turn off and its collector voltage to rise and produce a 1 for $\overline{Q}$. Since Q1 is off, output Q is now 0. Thus, conditions have gone from condition (1) to condition (2). To bring them back to condition (1), switch S is closed. Thus, the reaction of this circuit to a new pulse depends upon whether it is in condition (1) or condition (2) when the pulse arrives, and whether the pulse is applied to R or S. In practice, R and S inputs are never applied at the same time.

In practice, the switches at R and S are transistors. An example of how they can be connected is shown in Fig. 2-53(A). Inputs to switching transistors Q3 and Q4 can now be regular digital pulses rather than the short circuits of Fig. 2-52.

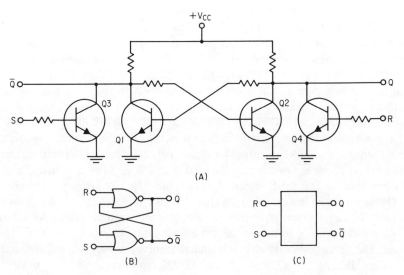

Fig. 2-53   *(A) Flip-flop with transistor switch, (B) logic circuit equivalent, and (C) symbols for an RS flip-flop.*

The flip-flop function can be shown as a combination two-input NOR gates with cross-connected outputs, as indicated in Fig. 2-53(B). By tracing the inputs and outputs through this logic, it can be seen that the function is the same as that indicated in Fig. 2-53(A). These diagrams are types that might be used where the contents of a flip-flop package are being discussed. In most logic diagrams, the flip-flop package is treated as an integral unit. The symbol for this RS flip-flop is illustrated in Fig. 2-53(C).

An RS flip-flop which is more useful in applications such as counters is the *RST* version, illustrated in Fig. 2-54(B). The added symbol *T* stands for *toggle*. It provides the means for triggering the added steering circuit (using Q5), which steers successive pulses coming in at T first to Q1, then Q2, then back to Q1, etc. As a simple example of what such a circuit can do, consider a problem in which we want to know whether an odd or even number of pulses has been fed into T. If we start with Q = 0 and $\overline{Q}$ = 1, then a 1 at Q will always indicate an odd number and at $\overline{Q}$ an even number. The symbol for this RST flip-flop is shown in Fig. 2-54(A).

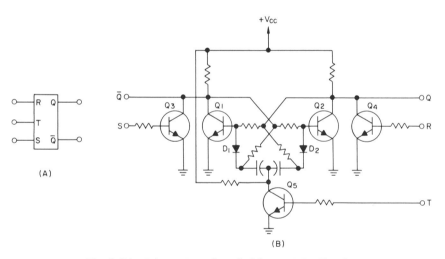

Fig. 2-54 *Schematic and symbol for an RST flip-flop.*

Symbols of other types of flip-flops are shown in Fig. 2-55. The clocked RS flip-flop at Fig. 2-55(A) is the same as the RS version, except that the operation is controlled by clock pulses, applied at C. The flip-flop action is initiated by the arrival of a clock pulse and is reinitiated with the advent of each new clock pulse.

In Fig. 2-55(B) is the symbol for a Type D flip-flop. This is also referred to as a *latch*. It is *"clocked"* only when a clock pulse arrives and the data at the input are transferred to the output (D stands for data).

In Fig. 2-55(C), is the symbol for the JK flip-flop. This is designed so that use can be made of the 1-1 input combination not used in the RS and RST

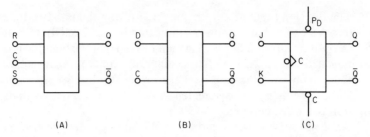

Fig. 2-55   Schematic symbols for three flip flops: (A) clocked RS, (B) type D, and (C) JK.

flip-flops. It uses fast negative-going pulses to raise inputs to level before clock pulses arrive. As will be covered in Chap. 9, flip-flops can be used in such things as counters, shift registers, etc.

Memory Devices

The flip-flops discussed above are a kind of memory device, in that each "remembers" what the previous input was. For permanent or semipermanent storage of data in a computer, some kind of matrix using either magnetic cores or solid-state devices is usually used. Core memories use arrays of small, doughnut-shaped pieces of magnetic material coupled to wires, through which electric pulses are sent to drive them to saturation (a stored 1) or read out from them whether a 1 or 0 is present. An example of the arrangement is shown in Fig. 2-56. Access to the information in a given bit of the memory is obtained by sensing its two coordinates (X and Y). The sense and inhibit lines are used for reading out and reading in information, respectively. Such memories are built into *planes* which are stacked one on another so as to minimize space occupied. Solid-state memories are made up of flip-flop circuits arranged in matrices like those of the core type.

Logic and memory packages are combined to form the larger modular parts of computers. How this is indicated in schematic diagrams is demonstrated in Chap. 9.

Integrated Circuits

In many circuits, there are physical building blocks of another concept—the *integrated circuit* (IC). An IC is the combination of components into a single physical entity. That entity is such that only it as a whole, and not any of its parts, is replaceable. This is because the ingredients that form the components and their interconnections combine with the base (called a "chip") on which they are mounted.

The most important advantage of ICs is their compactness: hundreds of components can be mounted on a chip having an area only a fraction of a square

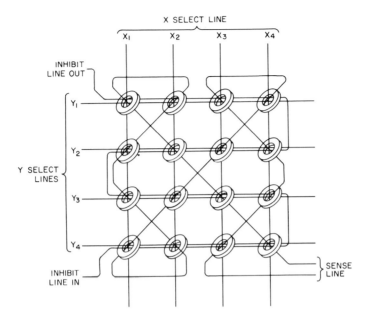

Fig. 2-56 *Schematic symbol for a core memory.*

inch. The second advantage of ICs is that manufacturers provide optimized circuit modules in IC form, making design of equipment easier and more efficient. Besides being convenient physical entities, ICs are also usually constructed so as to include some clearly recognizable functional part of an overall circuit (amplifier, oscillator, etc.). Thus, they are often helpful in analysis and in modular replacement in servicing.

We are interested here mainly in how ICs are represented on schematic diagrams and how to recognize and interpret their meaning. Therefore, we will consider a few examples of common types and the symbology involved.

An example of how a simple IC is constructed is shown in Fig. 2-57. The drawing in Fig. 2-57(A) shows how the added materials work with the chip to form a transistor, a resistor, a diode, and a capacitor. In Fig. 2-57(B) is the schematic diagram of the resulting circuit. Some ICs are not in themselves circuits, but merely arrays of components which can be interconnected by external wiring through connecting terminals. For example, there are arrays of transistors, diodes, resistors, etc. Available ICs then range up to types in which almost all circuits for a device (such as a small AM radio) are available on a single chip. It must be emphasized that Fig. 2-57 is a very much enlarged representation of the normal IC; such a circuit could fit on a chip only a tiny fraction of a square inch in size.

From the standpoint of electrical function, ICs may be divided into two types: linear and digital. The *linear* type is designed for applications where signal currents vary over a wide range of values and the output signal must follow the

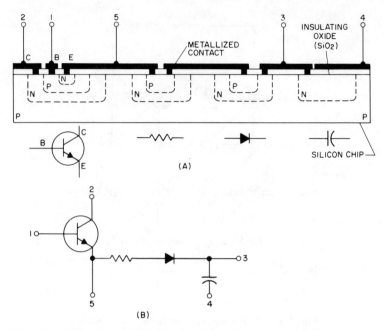

Fig. 2-57   *Drawings showing how a simple IC is constructed: (A) construction, and (B) schematic.*

input throughout this range. In other words, they handle signals with a continuous waveform. The *digital* type is for circuits that operate only on pulsed inputs and outputs and where the output needs only to respond either to full currènt or an "off" (or very low value of) current. Thus, digital ICs handle a discontinuous waveform.

Examples of linear IC applications: audio and RF amplifiers, oscillators, detectors for radio receivers, and most circuits for use in home radio, TV, and in much communications equipment. Digital ICs are mostly for computers and peripheral equipment, but are also widely used in control equipment switching. For example, a digital IC might include one or more AND or OR gates or flip-flops. Others group logic packages into such things as registers and counters (discussed in Chap. 9).

The schematic diagram usually makes no overt distinction between linear and digital ICs. The type of IC in question is usually suggested by the type of equipment under consideration, and by checking the number of the IC in the manufacturer's literature.

Perhaps the simplest ICs are those that are not in themselves a complete circuit, but instead provide *arrays* of components. Examples of these are shown in Fig. 2-58, (A) being a diode array and (B) being a transistor array. Notice that all terminals for the components are brought out to pins for connection to external components and power sources. This way, such an IC can be utilized in

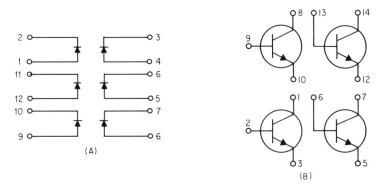

Fig. 2-58   Examples of IC arrays: (A) diode, and (B) transistors (Courtesy, RCA).

any of a very large number of ways by making external connections suit the purpose.

To show how such an array type IC might be used and diagrammed, consider Fig. 2-59. This shows how two transistors of the array of Fig. 2-58(B) might be used for the transistor-transistor logic (TTL) circuit of Fig. 2-48(E). The IC is shown as a triangle, with its external connections around the periphery numbered in accordance with the physical coding of the IC. Components other than the transistors (in this case only resistors) are shown external to the transistor as they are wired into the external circuit. Sometimes rectangles, rather than triangles, are used to depict ICs in schematic diagrams.

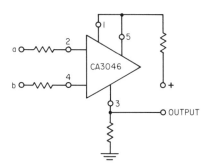

Fig. 2-59   RCA IC array connected into the circuit of Fig. 2-48(E).

Other than arrays, there are many ICs that have internal circuit connections and need only a minimum of external components and power from external circuits. The tendency is to use a widely adaptable circuit which can be connected in any of a wide variety of arrangements. Probably the two most widely used functional circuit IC arrangements are the *differential amplifier* and the *operational amplifier.*

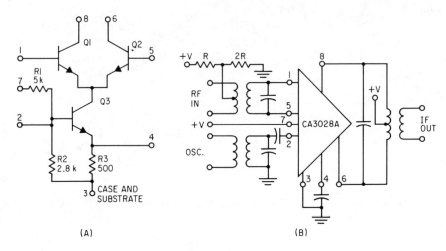

(A)   (B)

Fig. 2-60 *Differential amplifier RCA CA 3028 A: (A) circuit, and (B) use in mixer circuit (Courtesy,* RCA).

A typical differential amplifier IC is illustrated in Fig. 2-60(A). The emitters of transistors Q1 and Q2 are connected, and this junction is connected in turn to the collector of Q3. Q3 acts as a constant-current source for emitter current. Chips are available with two differential amplifiers. Besides their use as simple amplifiers, they will also work in such applications as oscillators, mixers, and frequency synthesizers. This IC is shown in a mixer circuit in Fig. 2-60(B).

The circuit of a typical operational amplifier (opamp) is shown in Fig. 2-61. It consists basically of two differential amplifiers connected in cascade with negative feedback. Such a circuit provides very high dc voltage gain, great bandwidth, high input impedance, and low output impedance.

We have considered only a few examples of the multitude of circuit arrangements available in ICs. The purpose here is to illustrate how these widely used devices are shown on schematic diagrams. Further examples appear in later chapters in specific diagrams.

Integrated circuits are produced in several physical forms, illustrated in Fig. 2-62. These are of particular interest because such representations are often included on schematic diagrams. In Fig. 2-62(A) is the T05 package, in which the chip is enclosed in a small (about one-third of an inch in diameter) can, with leads coming out of the bottom. In this case there are 10 leads, numbered counterclockwise from above, starting at the projection on the side. Then there are flat plastic packages as shown in Fig. 2-62(B). Again, the leads are numbered counterclockwise from the top, starting at top right, where there is a locating dot. A third, and probably the most popular type of IC package, is shown in Fig. 2-62(C). It is known as the dual-in-line package (DIP). This package, which comes in varieties having different numbers of pins, plugs into a socket, from which connections to the remainder of the circuit are made.

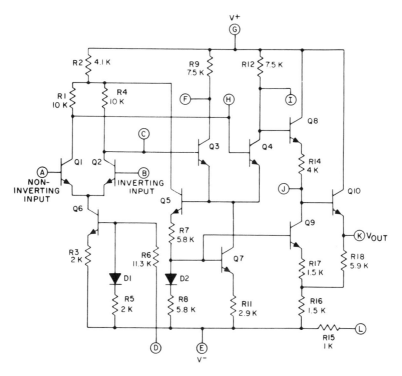

Fig. 2-61 *Typical opamp circuit* (*Courtesy,* RCA).

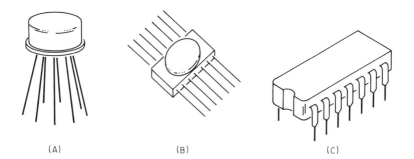

(A)   (B)   (C)

Fig. 2-62 *Three forms of IC packages: (A) TO5, (B) flatpack, and (C) DIP.*

CHAPTER

3
Functional Sequence
and
Block Diagrams

For any operating device, there is a *functional sequence* of events or actions. One action initiates others, which in turn initiate still others until the device has completed its overall function with some kind of output. In some devices, this sequence of functions is very obviously time-related. That is, the successive steps in the sequence are each long enough and distinct enough so that the individual part of the sequence is easily identifiable. An example of this is the home-type clothes washer. The fill, wash, rinse, and spin phases each take some time and their sequential arrangement is very clearly perceived. In fact, this sequence is usually marked on a timing dial.

On the other hand, consider a series of linkages, such as a pantograph, used to make a copy of a map or drawing by tracing the outline of an existing map or drawing. Here, it would seem that the response of the output motion is instantaneous, and the copy is executed at the same time as the original map or drawing outline is traced. For this reason, this is an excellent device to illustrate what we mean by functional sequence. Although the time elapsed between the initiation of motion and the response through the linkage to the copy is essentially zero, the action of each link of the device depends upon the action of the one that precedes it. Each link must receive force and motion from the preceding one before it functions, even though each function occurs at the same time. In other words, there is always a definite sequence in which parts of a device are "urged" into action. A clear knowledge of such a sequence is of critical importance in understanding and working with the device.

Electrical, and particularly electronic, devices are good examples of functional sequence because the time consumed in the propagation of a function through the device is, for everyday purposes, negligible, and everything operates (or seems to operate) at the same time. Thus, it is clear that by functional sequence we are not talking about quite the same thing as time sequence, even though they are usually coexistent. Of course, we know that a functional sequence can never progress backward in time; the important thing is that, for all practical purposes, each function in a sequence can be of near-zero duration (but never absolutely zero).

To become more specific in the electronics field, consider a radio receiver. The action in it is initiated by a *signal*, which is derived from an antenna (which may be a loop physically contained within the receiver). The signal is applied to the RF amplifier, whose output is applied to the mixer, then to the IF amplifier, etc. Each successive function in the chain depends on the receipt of a signal from the previous function. The IF amplifier needs a signal from the mixer, the mixer receives a signal from the RF amplifier, which gets it from the antenna. Although each function takes time, that time interval is very small, in the order of a millisecond and less, and, unless we are making extremely short time measurements, it is essentially zero. Thus, the important factor is the *dependency* of each successive function upon the preceding one.

Why Is Functional Sequence Important?

Schematic diagrams and other similar diagrams are laid out in functional sequence. Accurate interpretation of a diagram depends greatly on how easy it is to relate that diagram to the functions of the device which it represents. When we think of those functions, we must think mainly of the sequence of functions involved. Therefore, good schematic, block, and functional diagrams are drawn so that the functional flow is represented from *left* to right in the same order as the words in the sentences on this page. In fact, sentence word order is a good analogy. Consider the following sentence:

Important functional is sequence very.

This "sentence" is either unintelligible, or it is understandable only after some deliberation. To be easily read and understood, the work order must be corrected to:

Functional sequence is very important.

If we should draw a diagram of a washing machine with the spin-dry function at the left, the initial fill in the middle and the wash cycle at the right-hand side, we would have the same kind of problem reading this diagram as we did with the mixed-up "sentence" above. Even though we show the relationships by arrows connecting the symbols, the placement of these symbols on the diagram could still be confusing because they would not be in order of functional sequence. Thus, we can state a very important rule:

Rule: Wherever possible, schematic, block, and functional diagrams are laid out to represent functional flow from left to right.

The Block Diagram

When it is not necessary to show specific connections or to show the nature of each component, but it is desired to show functional scheme, a *block diagram* is in order. As the name implies, in block diagrams we use blocks, which may be rectangles (most common) or other geometric figures. Each block represents a functional unit of the circuit or device. The functional unit may be at any of several levels, from (in rare cases) a single component to a large circuit of many stages. In electronics, a block is most frequently a functional portion, such as an amplifier, oscillator, or detector, or a group of them.

The level of each block depends primarily on the level of the whole block diagram, as will be presently illustrated. For example, a block diagram of a radio receiver would show separate blocks for RF amplifier, mixer, and oscillator. On the other hand, a block diagram of a radio station installation in which there are a number of receivers and transmitters might show each receiver and each transmitter as a block.

If the basic circuit depicted by a block is repeated, sometimes a single block may include all the repetitions. For example, the IF amplifier section of a radio receiver frequently contains two or more IF amplifier stages. In that case, it is usual to represent both (or all) IF stages in one block, designated the "IF Amplifier."

Simple Example of Block Diagram

Now consider a schematic diagram of a very simple radio receiver and how it may be represented by a block diagram. The device depicted in Fig. 3-1(A) is a *crystal set* used with earphones. Its schematic diagram appears in Fig. 3-1(A). Component D1 is a solid-state rectifier diode whose schematic symbol is discussed in Chapter 2. The symbol PH is for earphones. This type of receiver cannot produce sufficient output signal to operate a loudspeaker (since there is no audio amplifier), so the earphones are appropriate.

One method of showing this device in block diagram form is illustrated in Fig. 3-1(B). The circuit itself is divided into two separate parts: the tuning circuit and the detector circuit. A block is assigned to each, with the earphones and the antenna included as separate functional units. Note here a very important distinction:

1. In the schematic diagram, a line between components represents a connection or a conductor.
2. In the block diagram, a line between blocks indicates functional flow.

Normally, a component requires at least two conductors to connect it into a circuit. Circuit sections often need more than this. Each of these connections must be shown in a schematic diagram. In the block diagram,

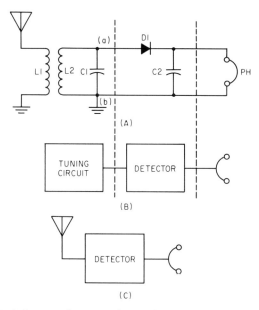

Fig. 3-1 *(A) Schematic diagram of a simple radio receiver, and (B) and (C) two ways of showing a block diagram.*

however, the important thing is the functional flow from one block to the next. In Fig. 3-1(A), for example, the schematic diagram shows the tuning circuit connecting to the detector through two conductors at (a) and (b). But since these connections are made to transfer a single signal from one section to the other, the block diagram shows only one line between them. Similarly, earphones require two connections, as shown in Fig. 3-1(A), but the functional signal flow to them in Fig. 3-1(B) is indicated by a single line.

The extent of the circuit represented by a block is not fixed or standardized but depends on what functional emphasis is to be given in a particular case. In Fig. 3-1, for instance, we might want to consider the tuning circuit as part of the detector circuit and show the block diagram as in Fig. 3-1(C). This is actually the more common way to show such a circuit in block form, unless there is something functionally special about the tuning circuit to require the emphasis of a separate block.

Other Block Diagrams

Suppose now that we make our receiver into a simple superheterodyne, whose circuit is shown in Fig. 3-2(A). This receiver has no RF amplifier but starts with a mixer (Q1); then, there is an oscillator (Q2), one stage of IF amplification (Q3), then a detector (D) and earphones. The block diagram of this receiver is shown in Fig. 3-2(B). The antenna is coupled to the mixer, as is the

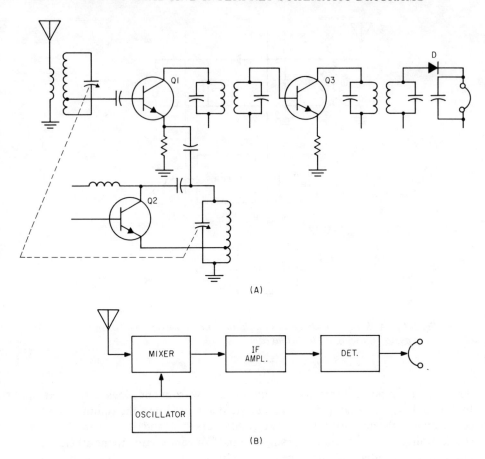

Fig. 3-2 *Simple superheterodyne receiver: (A) schematic diagram, and (B) block diagram.*

receiver's local oscillator. The frequency of the oscillator signal is such that it can heterodyne with the incoming signal and produce a signal at the intermediate frequency. The latter is amplified (and selected) in the IF amplifier and fed to the detector, which derives the modulation from the signal to make it audible in the earphones.

Notice that the oscillator block is placed *below* that of the mixer in the diagram. This results in functional flow from the oscillator going upward rather than from left to right. This is a variation from the left-to-right rule because the oscillator signal is really just joining an already established left-to-right stream. Often, it is desirable, for simplicity and clarity, to have such limited vertical functional flow. In some cases, even right-to-left functional flow may be used for a minor part of a diagram. An example of this is a feedback circuit, in which some of the output is fed back to the input. However, this does not alter the fundamental left-to-right flow, since it depicts *feedback*.

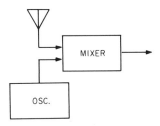

Fig. 3-3 *Different positions of the front-end blocks in the superheterodyne block diagram.*

Actually, there are a number of options for placing the oscillator in the diagram. One is illustrated in Fig. 3-3. The oscillator block is put a little to the left and below the mixer in such a way that the functional flow from the oscillator is primarily left-to-right. Flow from other sources outside the main stream of function are often shown in other than left-to-right (especially up or down) to suit convenience and clarity. Thus, the arrangements for the oscillator in Figs. 3-2 and 3-3 are equally acceptable.

Sometimes, the functional flow can be initiated by any of several sources. In that case, the block diagram may be arranged more like the one shown in Fig. 3-4. This depicts an audio system that can operate from microphones or a record player turntable. In the mixer, one signal is rejected or accepted, or the signals can be used in varying proportions. Notice the use of a large loudspeaker symbol for the *woofer* or base output and small symbol for the *tweeter*, or treble and middle range output.

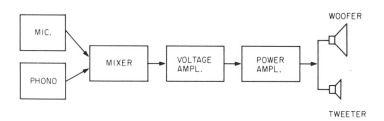

Fig. 3-4 *Arrangement of blocks where two or more input sources may be involved.*

For further illustration, consider the conventional superheterodyne radio receiver depicted by the block diagram of Fig. 3-5. Notice that one block is used for the IF and another for the audio frequency (AF) section, even though each may contain two or more amplifier stages. Of course, if there is something functionally important and different about one of the IF or AF stages, there is nothing to say we should not use a separate block for that stage.

The example in Fig. 3-6 is a block diagram showing the organization of a microcomputer system. Since this system comprises many large circuits, each block represents much more than in the diagram of Fig. 3-5.

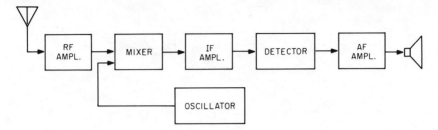

Fig. 3-5   Superheterodyne receiver block diagram.

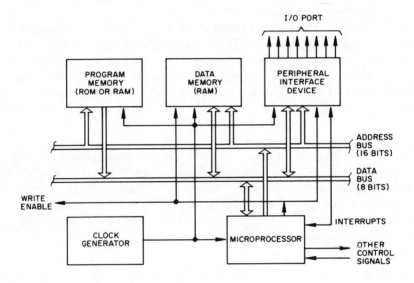

Fig. 3-6   Block diagram showing organization of a microcomputer system.

There are many applications of block diagrams, both in electronics and in other fields. We have used these simple examples merely to illustrate the general applications. More is discussed about various applications in the chapters that follow.

Wiring Diagrams

Wiring diagrams are those that show actual conductor interconnections between components or devices, with some relation to the physical layout. Such a diagram is designed to assist in wiring or tracing wiring, and it shows each conductor as it exists. For example, if two conductors both connect to the same two terminals, there is more than one way this may be accomplished. Two conductors can both go to one terminal, with a separate conductor connecting the two terminals. Or, leaving the two terminals connected, one might connect one conductor to one terminal and the other to the other.

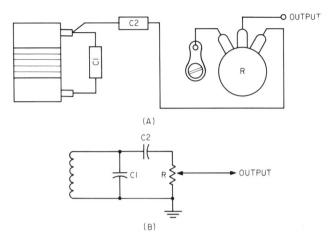

(A)

(B)

Fig. 3-7   *Simple example showing the difference between a wiring diagram (A) and a schematic diagram (B) for the same circuit.*

A very simple example illustrating the difference between a wiring diagram and a schematic diagram is shown in Fig. 3-7. Physical representations of the components are used, with emphasis on the location and arrangement of the electrical connections of each. Component symbols are, wherever possible, located approximately as they are in the equipment, so that the scheme of wiring can be accurately shown. Figure 3-7(A) shows the wiring diagram for a small portion of a circuit. The corresponding schematic diagram is shown in Fig. 3-7(B). Notice in Fig. 3-7(A) that not only are the connections to variable resistor R shown, but also which conductors must connect to which terminal. For such a component, if the connections are not properly made, the component will either not operate at all or may operate improperly.

Where a number of conductors must follow nearly the same path for some distance in the equipment, it is often desirable to provide *cabling*. The conductors are bunched together and either tied together with cord, or enclosed in sleeving that holds them together. Individual conductors can be brought out along the length of the cable, to run to their intermediate destinations. An example of a simple wiring diagram of a circuit in which cabling is used is shown in Fig. 3-8. Notice that the main cable, which travels down the center, "breaks out" into two parts at the top and that individual conductors break out on each side in the middle of the cable. Notice also how the conductor cabling is indicated by dashed lines circling the conductors involved. If the conductors are enclosed in a grounded metal shield, the dashed lines are connected to a lead to ground.

Cables within a device usually terminate in individual connections of the conductors. External cables (those which connect physically separate units) usually are fitted with *connectors*. Examples of some of the more common connectors and how they are shown on a schematic diagram are illustrated in

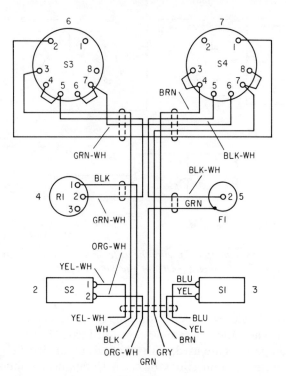

Fig. 3-8   *Wiring diagram, showing the use of a cable.*

Fig. 3-9. Generally, connectors are divisible into two groups: *plugs* and *jacks*. A plug, also known as a *male* connector, has a prong or other metal projection. The jack, or *female* connector, is designed to receive the prong or projection of the plug with which it mates. Connectors are used to connect cables with chassis or cabinets, cables with each other, or plug-in modules with the remainder of a piece of equipment.

   Whether or not a plug or jack is used in a given location is usually determined by whether the circuit to which it is attached is a source or user of power. If the cable is to provide electrical power to a device, the device end of the cable should have a jack; if a plug were used, short circuits or hazards would result from the projecting prong or prongs. For the same reason, the cable would have a plug at the power supply end, to mate with a jack on the power supply unit. Many connectors have only one prong (or one and a ground); others have two, three, or many more. Some are shielded so that they can be used with shielded wire or coaxial cable.

   Figure 3-9(A) illustrates what are probably the most widely known plugs and jacks: the plugs used on household appliances, which mate with the jacks which supply electric power in homes. These particular jacks are usually referred to as *receptacles*. Notice the two types of symbols for plugs, shown at the

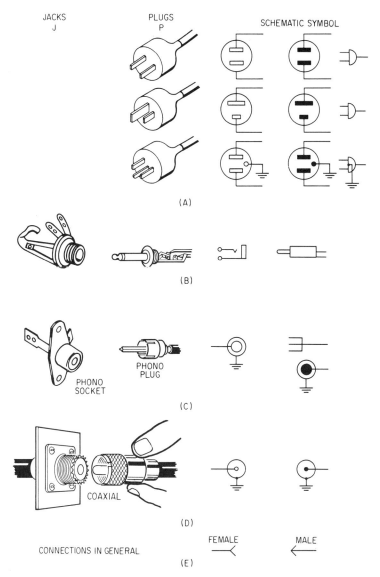

JACKS
J

PLUGS
P

SCHEMATIC SYMBOL

(A)

(B)

PHONO
SOCKET

PHONO
PLUG

(C)

COAXIAL

(D)

CONNECTIONS IN GENERAL

FEMALE

MALE

(E)

Fig. 3-9 *Symbols for cable connectors in schematic diagrams.*

far right. In the second row from the right the black rectangles indicate prongs, that is, plugs. In the third row from the right are the receptacles, or jacks. The white rectangles indicate holes 'and identify these as jacks.

In Fig. 3-9(B) is shown a *phone plug*, used primarily in AF applications, particularly for earphones and loudspeakers. It has two terminals, one of which may be grounded, but does not have to be. (Some plugs have more than two terminals.) Figure 3-9(C) and (D) show connectors for shielded or coaxial cable. Notice that the same type of code to distinguish between plug and jack is used as in Fig. 3-9(A). The black circular area in the center represents a plug, and the white area, the jack. In circuit diagrams a number of other types of connectors may be used. Schematic representation is usually like that shown in Fig. 3-9(E). Such symbols may be arranged in groups to represent multi-pronged connectors popular for connecting different sections of a piece of equipment together electrically.

Interconnect and Functional Block Diagrams

Interconnection diagrams are wiring diagrams showing how complete units or modules are connected together. For this reason, they usually involve cabling. A very simple interconnect diagram is shown in Fig. 3-10.

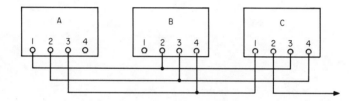

Fig. 3-10   *Simple interconnect type of diagram.*

Functional block diagrams are block diagrams with emphasis on a particular functional path through the equipment. An example appears in Fig. 3-11. This type of diagram is particularly useful for servicing equipment and often is found in service manuals.

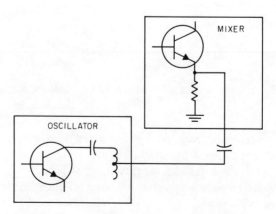

Fig. 3-11   *Form of functional block diagram.*

Printed Circuit Boards

In today's electronic equipment, the most common type of wiring is that of a *circuit board*. A circuit board consists of insulating board clad with copper on one side; the "wiring" is formed by etching away all the copper except that needed for conductors and terminals. A typical circuit board layout is illustrated in Fig. 3-12. The leads from components to be connected are pushed through terminal holes in the board and soldered there. The connections among these components are completed by the unetched copper areas running between them.

Notice in Fig. 3-12 that the components are added in dashed lines. The dashed lines indicate that the copper connecting material and the components are on opposite faces of the board. A piece of equipment may contain one, two, or many of these boards. An interconnect diagram showing cable connections to the boards would then complete the wiring information.

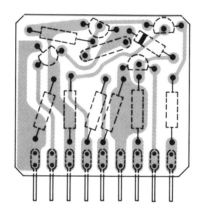

Fig. 3-12   *Typical circuit board layout.*

Physical Layout and Location Diagrams

Physical layout diagrams primarily show the locations of components in a device. This facilitates servicing or checking, since to find a given component one might have to check through much of the circuit. Figure 3-13 is an example of a physical layout diagram. Where printed circuit boards are used, the wiring diagram often serves also as a component location diagram.

Labeling and Parts Lists

As various schematic symbols have been introduced, we have mentioned the typical letter symbol for each component. For example, the letter symbol R stands for a resistor, L for an inductor, etc. Every resistor on a diagram is identified by the same letter—R. Resistors are distinguished from each other

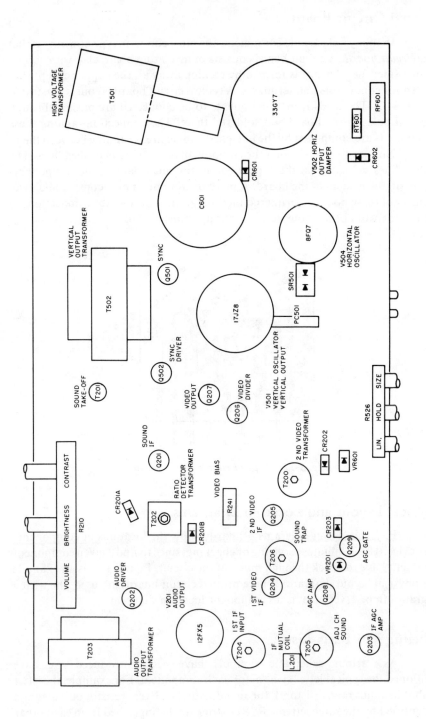

Fig. 3-13   *Physical layout diagram.*

by the addition of a number to the letter R. For example, the first resistor labeled might be called R1, the second R2, etc. Similarly, inductors may be labeled L1, L2, etc., capacitors C1, C2, etc., and so on. Thus, each component has a unique designation. If we want to give the specifications for any component on a list of parts, we can identify it with its letter and reference number.

In using a schematic diagram it is often very important that we be informed of the specifications of any or all of the components in it. This is the job of the *parts list.* Figure 3-14(A) is a simple illustration of a parts list. Notice the types of information given for each component. Usually of prime interest is the *value,* that is, resistance of a resistor, capacitance of a capacitor, inductance of an inductor, etc. Then comes other information, such as capacitor type (mica, paper, electrolytic), resistor power rating, and sometimes whether it is a composition or wirewound type.

A parts list is not always used. Sometimes it is more convenient simply to include the information with the component symbols on the diagram as illustrated in Fig. 3-14(B). Often, the information with each label is limited, because there is a general statement on the diagram, such as: "All resistors are ½-W composition type unless otherwise indicated."

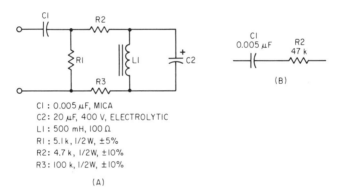

Fig. 3-14 *Simple circuit with (A) a typical parts list, and (B) the manner in which component information is sometimes placed on the diagram when a parts list is not considered necessary.*

It is of some interest in schematic diagram labeling to note that there are standard values of resistance and capacitance for commercially available components. Standard resistance values are listed in Fig. 3-15. The same list also applies to capacitor values in picofarads.

Values in component labels cover a very wide range. For this reason, not only are standard units used, but also multiple and submultiple units. The standard units for resistance, capacitance, and inductance are the ohm, the farad, and the henry, respectively. The most common multiples and submultiples are as follows:

$1\,M\Omega = 1$ megohm $= 1000\,k\Omega = 10^6\Omega = 10^6$ ohms
$1\,k\Omega = 1$ kilohm $= 1000\,\Omega = 1000$ ohms
$1F = 1$ farad $= 10^6\,\mu F = 10^{12}$ pF
$1\,\mu F = 1$ microfarad $= 10^6$ pF $= 10^6$ picofarads
$1H = 1$ henry $= 1000\,mH$
$1mH = 1$ millihenry $= 1000\,\mu H = 1000$ microhenries

STANDARD
COMPONENT VALUES

20% Tolerance	10% Tolerance	5% Tolerance
10	10	10
		11
	12	12
		13
15	15	15
		16
	18	18
		20
22	22	22
		24
	27	27
		30
33	33	33
		36
	39	39
		43
47	47	47
		51
	56	56
		62
68	68	68
		75
	82	82
		91

Ref. No.	OHMS						MEGOHMS		
	0 to 1.0	1 to 9.1	10 to 91	100 to 910	1,000 to 9,100	10,000 to 91,000	0.1 to 0.91	1.0 to 9.1	10 to 91
10	—	1.0	10	100	1,000	10,000	0.1	1.0	10.0
11	—	1.1	11	110	1,100	11,000	0.11	1.1	11.0
12	—	1.2	12	120	1,200	12,000	0.12	1.2	12.0
13	—	1.3	13	130	1,300	13,000	0.13	1.3	13.0
15	—	1.5	15	150	1,500	15,000	0.15	1.5	15.0
16	—	1.6	16	160	1,600	16,000	0.16	1.6	16.0
18	—	1.8	18	180	1,800	18,000	0.18	1.8	18.0
20	—	2.0	20	200	2,000	20,000	0.20	2.0	20.0
22	—	2.2	22	220	2,200	22,000	0.22	2.2	22.0
24	0.24	2.4	24	240	2,400	24,000	0.24	2.4	24.0
27	0.27	2.7	27	270	2,700	27,000	0.27	2.7	27.0
30	0.30	3.0	30	300	3,000	30,000	0.30	3.0	30.0
33	0.33	3.3	33	330	3,300	33,000	0.33	3.3	33.0
36	0.36	3.6	36	360	3,600	36,000	0.36	3.6	36.0
39	0.39	3.9	39	390	3,900	39,000	0.39	3.9	39.0
43	0.43	4.3	43	430	4,300	43,000	0.43	4.3	43.0
47	0.47	4.7	47	470	4,700	47,000	0.47	4.7	47.0
51	0.51	5.1	51	510	5,100	51,000	0.51	5.1	51.0
56	0.56	5.6	56	560	5,600	56,000	0.56	5.6	56.0
62	0.62	6.2	62	620	6,200	62,000	0.62	6.2	62.0
68	0.68	6.8	68	680	6,800	68,000	0.68	6.8	68.0
75	0.75	7.5	75	750	7,500	75,000	0.75	7.5	75.0
82	0.82	8.2	82	820	8,200	82,000	0.82	8.2	82.0
91	0.91	9.1	91	910	9,100	91,000	0.91	9.1	91.0

Fig. 3-15 Standard resistance values.

For the labeling of resistor power ratings and other information sometimes included on schematic diagrams, the following power units are encountered:

$1\,kW = 1$ kilowatt $= 1000$ W
$1\,W = 1$ watt $= 1000\,mW = 1000$ milliwatts

Sometimes, particularly to help in servicing, labels giving normal voltages and currents at particular points in a circuit are included. Commonly used multiples and submultiples of the volt and the ampere are as follows:

$1\,kV = 1$ kilovolt $= 1000$ V
$1\,V = 1$ volt $= 1000\,mV$
$1\,mV = 1$ millivolt $= 1000\,\mu V = 1000$ microvolts

1 V $= 10^6 \ \mu V$
1A   = 1 ampere   = 1000 mA
1 mA = 1 milliampere = 1000 μA = 1000 microamperes
1 A $= 10^6 \ \mu A$

We shall not dwell on parts lists and exact component values, since they are not at the heart of schematic diagram interpretation. However, it is true that the value or rating of one or more of the components in a diagram can affect the function in some cases. The principles of labeling and parts listing should, therefore, be kept in mind, even though they receive little more direct attention in the remainder of the book.

4
Power Supplies: Active Devices and the Need for a Power Supply

The design and study of an electrical device primarily involves the function of, or signal flow in, that device. Up to this point, we have focused on these concepts. In the real world, however, there is also another factor we must consider: any active device must be powered, that is, power from a source not necessarily related to the signal must be provided.

An *active* device is one which adds something to circuit operation; for example, a device which increases the level of the applied signal and thus puts out more signal than it takes in. By the fundamental law of the conservation of energy, the added power must be supplied from some additional source. This source is usually referred to as a *power supply*. The power supply provides direct current (most often) or alternating current which the device it powers uses to accomplish its function.

The opposite of an active device is a *passive* device, which requires no power other than that derived from signal sources. A passive device controls or modifies the signal it receives so that it will have some new desired form or magnitude.

Most prominent among active devices are transistors and vacuum tubes. A transistor can amplify a signal, but to do so, must have potentials applied to the elements and be supplied with *bias currents* which flow through these elements.

Very simple examples of passive (A) and active (B) devices are shown in Fig. 4-1. A transformer, a variable capacitor, a fixed capacitor, and a resistor are included in the circuit in Fig. 4-1(A). Each of these is a passive component, so the whole circuit is passive. The resistor absorbs power, but cannot increase it. There is no way to build up signal power in this circuit except to increase the input signal level. Thus, signal power cannot be amplified in this circuit but can be reduced by the resistor and the resistance in the transformer (also by resistance in the capacitors which in good components is negligible). Since the circuit in Fig. 4-1(A) cannot increase signal power, it must be classified passive.

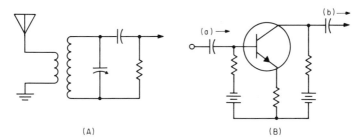

Fig. 4-1 *Examples of (A) passive, and (B) active circuits.*

In Fig. 4-1(B) the situation is different. One active component, the transistor, is used; thus, this is an active circuit. The signal enters the circuit at (a) and leaves in amplified form at (b). In a way, the circuit may be thought of as a *converter*, because some of the power fed from the batteries in the form of direct current is converted into additional signal power.

The importance of the above example is that no active device can operate without a power supply. Since most useful electronic devices are active, power supplies are extremely important.

Types of Power Supplies

A wide variety of power supplies is required for different needs. These can be divided into the following categories:

1. Batteries (chemical sources) and fuel cells
2. Electromechanical generators
3. Ac-to-dc supplies
4. Inverters (dc-to-dc and dc-to-ac)
5. Ac supplies (ac-to-ac)

It is desirable to review very briefly the purpose and nature of each of these types, because such knowledge is quite helpful in interpreting schematic diagrams.

Batteries are widely familiar because of their universal use in flashlights, radios, toys, and "cordless" appliances. Fuel cells are so called because they act as batteries that need to be continuously fed fuel as energy is drawn from them. Fuel cells are not in general consumer use but are limited to special applications such as spacecraft; thus, we shall not go further into these devices.

Electromechanical generating devices and their symbols are discussed in Chap. 2.

The most common supplies are probably ac-to-dc since there is at least one in almost every radio and TV receiver (except those operated only on batteries). The supply derives its power from the regular 117-V ac mains of house wiring (or, otherwise 230 V ac or other ac source). In many supplies, the voltage is changed to some desired level, then rectified and filtered for use in the device being powered. In cases where a change in voltage is not necessary, direct rectification without a transformer, followed by filtering, may be used. As will be seen, in some cases, voltage-multiplying circuits with two or more rectifiers and large filter capacitances are used to bring the input voltage up to desired dc output voltage. Three-phase and multiple-phase circuits are sometimes used with ac-to-dc supplies.

Inverters are used to change current from a dc source (such as a battery) into direct current at some other desired voltages. The approach is to use a *chopper* to turn the source current on and off rapidly so that the resulting varying current in a transformer primary will induce an ac voltage of a different desired value in the secondary. This voltage is applied to a rectifier-filter circuit to produce the desired dc output.

This scheme was at one time widely used in automobile radios. The chopping was originally accomplished by means of an electromagnetic vibrator, operating like a doorbell buzzer. Today, the same function is accomplished with much greater efficiency by transistor oscillators. A pair of transistors oscillate, that is, alternately turn the current on and off to produce the necessary alternating current for operating a transformer. In spite of this improvement, this system is no longer used in automobile radios, because the use of transistors and other solid-state devices allows direct use of the voltage from the 12-V automobile battery. However, this chopper technique is still employed in some special equipment.

The oscillating transistor arrangement is used in converters designed to produce an ac output from batteries, to operate electric shavers or other small appliances from the car battery, or from other battery sources. Here, the "chopped" alternating current is not rectified, but used as ac power for the device. In a few cases, an ac source is to be used for an ac device and the power supply merely raises or lowers the voltage (by means of a transformer) and controls it or adjusts it.

We will now consider schematic diagrams for simple examples of these different types of power supplies.

Batteries and Electromechanical Generators

Battery power supplies are available in several forms. Most common are 1) dry cells used in radio and TV receivers and in cordless appliances, and 2) storage batteries in automobiles. Many of the smaller batteries in radios and appliances are rechargeable.

The schematic diagrams for batteries and electromechanical generators are discussed in Chap. 2. Chargers for rechargeable batteries are discussed later in this chapter.

AC-to-DC Suppliers: Rectifiers

These supplies are universally used in radio and TV sets in the home and in many other electronic devices. The simplest type of ac-to-dc supply using a transformer is illustrated in Fig. 4-2(A). The standard 117-V, 60-Hz power source connects to the primary winding of the transformer, which steps the voltage up or down, depending on the level of dc output voltage desired. The ac secondary voltage is applied to rectifier D and the load in series. The rectifier passes current only on one "sense"[1] (electron current opposed to the arrow) so that the current through R and the load is direct. This current, and the voltage drop across R and the load are, of course, pulsating. In some applications, a pulsating output is acceptable, and this type of supply is sometimes encountered. The rectifier is polarized, that is, it must be connected into the circuit to yield direct current and dc voltages of the desired polarity. Figure 4-3 shows ways in which diodes are often marked to indicate polarity. In any event, whatever type of diode is used, it must be connected with proper polarity and the symbol on the schematic diagram shows how.

As mentioned before, the rectifier passes current only in one sense. As a result, only the positive half-cycles of the ac input appear in the output dc voltage. This is illustrated by the waveforms above the diagrams in Fig. 4-2.

---

[1] The term "direction" is often used here instead of "sense." In the case of a discrete conductor, however, current can be considered only in either of two senses. In a horizontal conductor viewed from the side, it can be only left or right, in a vertical conductor, up or down, etc. Thus, sense is in this case more appropriate.

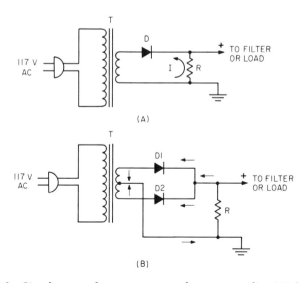

Fig. 4-2   Simplest transformer type ac-to-dc power supplies: (A), half-wave rectifier, and (B) full-wave rectifier.

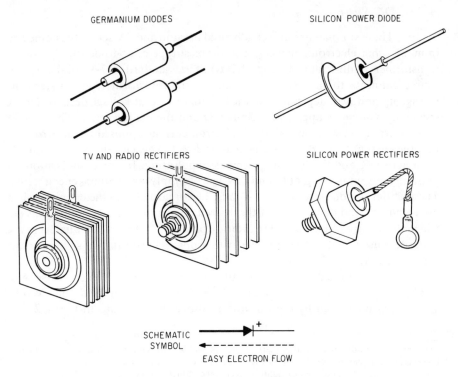

GERMANIUM DIODES

SILICON POWER DIODE

TV AND RADIO RECTIFIERS

SILICON POWER RECTIFIERS

SCHEMATIC
SYMBOL

EASY ELECTRON FLOW

Fig. 4-3   *Some types of diodes, with markings to indicate polarity.*

To make filtering of the output easier, a full-wave rectifier, shown in Fig. 4-2(B), is frequently used. Important distinguishing features of this type of rectifier are that there are at least two diodes and the transformer secondary winding must be center-tapped. The secondary must have twice as many turns as the transformer used with the half-wave rectifier, shown in Fig. 4-2(A), for the same dc output voltage. In other words, there must be as many turns on each side of the centertap in Fig. 4-2(B) as there are in the whole secondary winding in (A), for the same output voltage.

The output of a full-wave rectifier is easier to filter than that of a half-wave rectifier because both half-cycles of the input ac voltage appear in the full-wave output. As indicated by the waveforms in Fig. 4-2, the negative half-cycles are effectively "flipped over" so that the half-cycles in the output direct current are adjacent to each other.

The number of half-cycles per second in the dc output is known as the *ripple frequency*. In the half-wave case, one half of each ac cycle appears in the output. This half-cycle, plus the zero period that follows it, constitutes one cycle of the dc ripple. Assuming this supply operates at the standard (US) ac power frequency of 60 Hz, the ripple frequency is also 60 Hz.

In the full-wave rectifier in Fig. 4-2(B), both half-cycles of each ac cycle appear in the dc output. Since each half-cycle is the same as each other half cycle, and since there are twice as many half-cycles in the full-wave output, the ripple frequency is twice that of the ac input, or 120 Hz. At the higher ripple frequency, smaller components can be used in the filter circuit that follows.

Full-wave rectification without the larger, center-tapped secondary winding is possible with a bridge rectifier, whose basic schematic diagram is shown in Fig. 4-4. In most cases (but not always), a bridge rectifier can be recognized by the diamond-shaped arrangement of the four rectifiers, D1 through D4. Occasionally a diagram with a different rectifier configuration will be encountered; an example is shown in Fig. 4-4(B). Note that electrically this circuit is identical to that in Fig. 4-4(A).

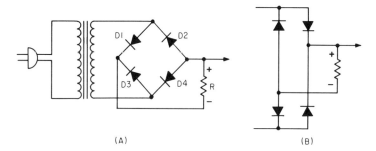

Fig. 4-4   *Two electrically identical diagrammatic arrangements for a bridge rectifier.*

To make sure that a given diagram does represent a bridge rectifier, a good method is to trace the path of current for each polarity of the source, as shown in Fig. 4-5. Here, the sense of the *electron* current when point *a* is negative (with respect to *b*) is shown by a solid-line arrow; electron flow while *a* is positive (with respect to *b*) is shown by a dashed line. The electron current, of course,

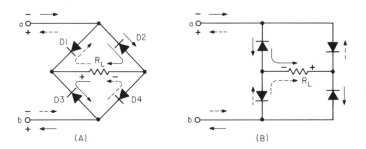

Fig. 4-5   *Electron current direction through a full-wave (bridge) rectifier and its components. Solid line arrows show current sense when a is negative; broken-line arrows, when a is positive. (B) is another way of drawing the same circuit as in (A).*

always has a sense opposite to that of the arrows of the diodes. For proper rectification, the sense of the current through load resistor RL should be the same, regardless of the source polarity. Because there is conduction through two of the diodes during each half cycle of the source voltage, this is a *full-wave rectifier*. In fact, this method can be used to analyze any given rectifier circuit. Note the sense of each diode (indicated by its arrow symbol) and then determine whether or not there can be current through the load during only one half cycle (polarity of the source) or during both half cycles of the source current.

Multiple-Phase Circuits

The ac sources discussed above are all what are called *single-phase* sources. There are two source terminals and the voltage varies from a maximum (peak) value in one sense downward through zero and to a maximum value in the other sense.

Let us now consider a type of source in which there is more than one voltage. In such a source, each voltage is called a phase voltage, because its timing, or phase, is different from the phases of other voltages in the source. Probably the most commonly used is the source known as the *Edison three-wire system*. This system has two phases and is universally used in electric current sources in homes. We will introduce the *3-phase* system first.

The 3-phase principle can be visualized by considering a simple generator, represented schematically in Fig. 4-6. If the generated voltages from

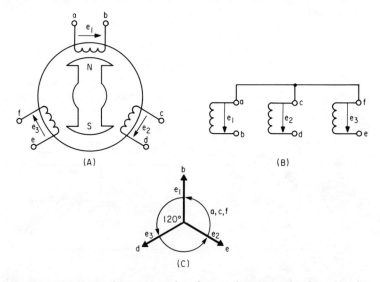

Fig. 4-6   *The three-phase principle: phase voltages can be thought of as generated by the permanent magnet rotating past the three coils in (A), which are connected as in (B) to produce the Y-connected configuration shown vectorially in (C).*

the three coils are connected as shown in Fig. 4-6(B), the vectors for the three voltages will form what is known as a *Y-connected* 3-phase source, in accordance with the resemblance of the configuration to a *Y* (upside down in this example) illustrated in Fig. 4-6(C).

The Y configuration is not the only one for 3-phase circuits. Another is the delta system, illustrated in Fig. 4-7.

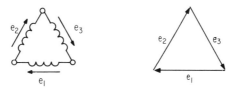

Fig. 4-7 *Coil connection and vectorial representation of a delta-connected three-phase system.*

If energy in a 3-phase system is to be transmitted over a transmission line (or just over local wiring) at least three conductors must be used. We say at least three because in a Y circuit the *neutral* (centerpoint) connection is often carried. The resulting circuit is known as the *4-wire* 3-phase Y system, illustrated in Fig. 4-8. Notice that, because there is no *common* point in the delta system, transmission with a neutral is not possible. Each voltage in a 3-phase system is separated in phase by 120° from each of the other two voltages.

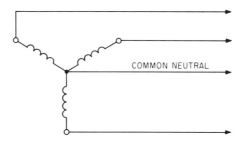

Fig. 4-8 *Four-wire, three-phase transmission.*

Let us now turn back again to the Edison 3-wire system. The source for this, as indicated in Fig. 4-9, is two voltages 180° apart in phase. This would appear to be a *true 2-phase system*[2] *because there are* two equally spaced vectors per cycle. But, in the Edison system, the two voltages are connected additively, so that a 3-wire system is formed (see Fig. 4-10). The common connecting point is the neutral wire. The voltage between conductor one and neutral is equal to

---

[2] However, what is called a 2-phase circuit employs two voltages 90° apart. It is rare and does not merit discussion here.

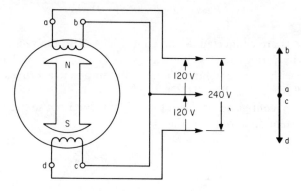

Fig. 4-9 *Edison three-wire system.*

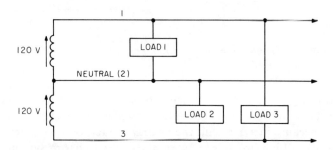

Fig. 4-10 *Load arrangements for Edison three-wire system.*

that between conductor three and neutral (normally 117 V in US homes). Loads can be connected between conductor 1 and neutral (load 1) or between conductor 3 and neutral (load 2). There is also the option of connecting a load (load 3) between conductors 1 and 3, which supply a voltage of twice either of the others (i.e., their sum). This allows for use of either 117-V or 235-V loads in the same system.

There are also systems having more phases than three. The separation in phase between successive voltages, for three and more phases is:

$$\phi = \frac{360}{n}$$

where ϕ is the separation in degrees and n is the number of phases. This information is of interest because, as in the case of the 3-phase circuit, each source voltage symbol is usually shown geometrically spaced at an angle having the same number of degrees as its phase separation. Since multiple-phase circuits of more than three phases in power supplies are relatively rare, we shall not discuss them further.

Power Supply Filters

As previously indicated, the output of a rectifier is pulsating direct current. In most applications, such current cannot be used directly but must first be filtered. In other words, the fluctuations of the current must be smoothed. This requires some sort of *filter circuit.*

The filter circuit in the schematic diagram of a power supply can be recognized by the presence of filter capacitors, choke coils (filter inductors), and resistors. Filter capacitors for power frequencies (that is, 60 Hz, 25 Hz, or 50 Hz) are high-value types and are usually electrolytics, which appear on the schematic diagram as polarized capacitors with the positive terminal usually marked with a "+." For the same power frequencies, filter chokes ordinarily are of 1 or more henrys, have iron cores and are heavy (the weight, of course, depending upon current rating). Filter chokes usually can be recognized in a diagram by the symbol for the iron core. Filter circuit resistors appear the same as any other resistors in the diagram, and so must be identified by their connection in the filter circuit.

The simplest type of filter circuit consists of only a single filter capacitor, as shown in Fig. 4-11(A). The circuit also shows R_B, known as a *bleeder* resistor. The half-cycle pulses from the rectifier charge C1 up to peak value. If the current output is not too great relative to the input to C1, the capacitor charge holds near the peak voltage, thus maintaining a nearly steady voltage output. From a steady-state analysis viewpoint, C1 can be considered to have a much higher dc resistance (to ground) than the load but much lower reactance than the impedance of the load. Therefore, most of the alternating component of the rectifier output current passes through C1, while the direct portion goes to the load.

Bleeder resistor R_B has two purposes. First, it maintains a steady minimum current from C1, thus avoiding the high voltage fluctuation that would otherwise take place if the load were suddenly removed. Second, it makes the power supply safer, because when it is turned off, R_B discharges the filter capacitor.

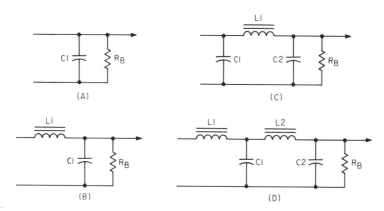

Fig. 4-11 *Filter circuits: (A) single capacitor with bleeder, (B) input choke added, (C) two-section with capacitor input, and (D) two-section with choke input.*

Now consider the other filter circuits in Fig. 4-11. In (B), filter inductor L1 has been added. A good inductor has a high reactance to minimize the alternating current and a low dc resistance to allow for as much direct current as possible. The reactance of the capacitor is low, so the already lowered alternating current produces only a low voltage drop across it. Thus, the ripple is greatly reduced in the output.

In Fig. 4-11(C), another capacitor is added before the inductor, for greater filtering. Because the first thing the output of the rectifier encounters is a capacitor, this is called a *capacitor-input* filter. By the same token, the arrangement in Fig. 4-11(B) is called a *choke-input* filter. The circuit of Fig. 4-11(D) represents the extension of (C) to become a choke-input filter by the addition of input choke L1. Choke input is most likely used in supplies where maximum current drain to the load is high and in which that current drain varies widely during operation. An example is a high-powered class B amplifier for an audio or modulated RF signal, in which the power source current must vary from near zero to full value as the AF signal fluctuates.

The choke used in a choke-input supply for such an amplifier is sometimes a *swinging* choke, a filter inductor in which the gap usually left in the laminations is omitted so that the core tends to saturate and reduce reactance at high currents, allowing more current through (although, of course, with somewhat less filtering). There is nothing about the symbol for a filter choke to indicate whether it is a swinging type or not, although occasionally the label *SW* is placed near it. Because of the wider use of voltage and current regulators in today's supplies, the swinging choke is not as common as it once was.

Voltage Multipliers

Sometimes, in order to eliminate the cost and bulk of a power transformer, power supply designers use only rectifiers and filter capacitors in a *voltage-multiplier circuit*. When tubes were used for radio and TV sets, such a multiplier circuit was often found. Plate voltages of 200- to 400-V dc were required, and the 117-V power line potential could be doubled or tripled to obtain this voltage without use of a power transformer. Today, with solid-state components having replaced tubes, and due to the lower voltages (say 12 to 150 V) required, multipliers are less often used in power supplies. However, they are still encountered in some TV circuits, particularly in TV high-voltage supplies.

Probably the simplest multiplier circuit is a half-wave doubler, illustrated in Fig. 4-12(A). It operates as follows: when point *a* is positive with respect to *b*, D1 conducts and C1 charges to the peak input voltage of the polarity indicated. When the input polarity reverses, D1 shuts off, and D2 conducts. The voltage on D2 consists of input dc voltage from *b*, plus the input peak (dc) voltage on C1. These cause current through the circuit D2-C 2-*a*-*b*-C1, charging C2 to twice the peak input voltage, and this "double" voltage is the

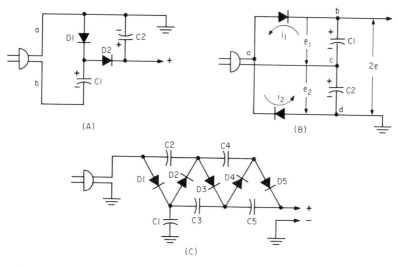

Fig. 4-12 *Voltage multipliers: (A) half-wave doubler, (B) full-wave doubler, and (C) quadrupler.*

output. This is called a *half-wave voltage doubler* because energy is derived from the source only when point *a* is positive with respect to point *b*.

Another doubler, referred to as a *full-wave voltage doubler*, is illustrated in Fig. 4-12(B). When point *a* of the source is positive, there is current i_1 through c-C1-b-a. This charges C1 as indicated. When c is positive, current i_2 through a-b-C2-c charges C2 as indicated. The voltages resulting from the charges on the capacitors add together to produce total output voltage 2*e* as shown.

Although the above examples are doublers, similar principles apply to circuits designed to multiply the source voltage by three or more. An example of an arrangement for multiplying by four is shown in Fig. 4-12(C). The principles involved are the same as those for the doublers previously discussed. The schematic diagram can be recognized by tracing through each diode, and determining the voltage developed across each capacitor.

Regulators

Modern power supplies, especially those for solid-state equipment, must produce voltages and/or currents that are essentially constant. For batteries this is not a problem, since battery voltage is usually quite constant (in the absence of overload), except for the drop occurring as the cells approach the end of their useful lives. Primarily we are concerned here with regulation in ac-powered dc supplies.

Perhaps the simplest of voltage regulators is the *nonlinear* type, illustrated by the circuits in Fig. 4-13. The one at (A) uses a gas tube, such as the

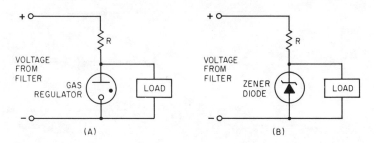

Fig. 4-13   *Voltage regulators: (A) gas tube, and (B) Zener diode.*

VR75, VR105, or VR150. Each of these tubes maintains a constant voltage (75, 105 or 150 V, respectively) across it even while the current through it varies over a wide range, like 5–30 mA. The value of resistor R is chosen so that the tube carries nearly its maximum current when there is no load. Application of a load draws some of the current away from the regulator without changing the voltage. Notice the symbol for the gas-tube regulator. Gas-tube regulators are seldom found in new equipment. They have been replaced by *Zener diodes*— special purpose diodes whose characteristics are similar to gas tubes' but which are physically much smaller. Zener diodes are available in a wide range of voltage and power ratings. But the power ratings are seldom shown on schematic diagrams.

A more efficient way of regulating output voltage is to put a transistor in series with the output and connect the transistor so that output voltage changes control the bias applied to it. Figure 4-14 is a schematic diagram showing how this can be done. In this case, a Zener diode sets a reference voltage which keeps the base of the transistor at a constant potential. Current from terminal *b* must pass through the transistor (collector to emitter) to go to the load. If the output (load) voltage goes up (more negative) the negative bias on the emitter goes up and the collector-to-emitter resistance goes up. This increases the resistance to load current and reduces the load voltage to compensate for the original increase. Similarly, if the load voltage goes down, the transistor resistance goes down and allows the load voltage to rise to compensate. Because the regulating

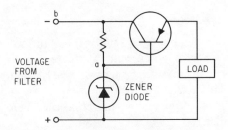

Fig. 4-14   *Transistor series regulator with base voltage controlled by a Zener diode.*

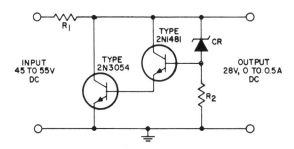

Fig. 4-15 *Shunt regulator (Courtesy,* RCA).

transistor acts in series with the load current, this type is often referred to as a *series regulator*, in contrast to the gas-tube and Zener diode circuits of Figs. 4-13 and 4-14, which are *shunt regulators*.

An example of a shunt regulator using two transistors is shown in Fig. 4-15. The Zener diode, CR, maintains a fixed voltage drop, so that when the output voltage tends to change, it causes a change of voltage drop across R2. Since the latter is the forward-biased base voltage of the 2N1481 transistor, this transistor conducts more, causing the other transistor to conduct more. The second transistor pulls extra current through R1, dropping the output voltage to compensate for its tendency to rise.

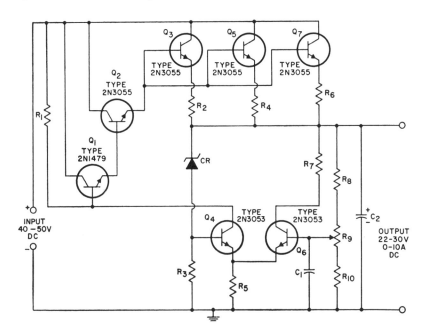

Fig. 4-16 *Series regulator using paralleled transistors and amplification of voltage variations (Courtesy,* RCA).

The degree of regulation is improved if even the slightest output voltage increase or decrease is amplified before being applied to the controlling element. An example of this technique is shown by the circuit of Fig. 4-16, which also illustrates use of several transistors in parallel, to increase current output capacity. The series regulation is accomplished by three transistors (Q3, Q5, and Q7) in parallel. These transistors are controlled, through Q1 and Q2, by Q4. The base voltage of Q4, derived from R3, varies with output voltage, since it is in series with constant-voltage Zener diode CR. The principle is the same as in Fig. 4-15. But, in addition, there is an adjustment for the output voltage at which regulation takes place. This is accomplished by R9, which varies the output

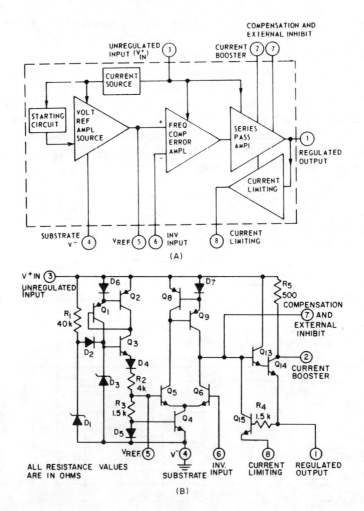

Fig. 4-17 *Block and schematic diagrams of an integrated circuit voltage regulator (Courtesy,* RCA).

voltage appearing at the base of Q6, so that, through Q6, Q4 will operate accordingly.

Regulation circuits for power supplies frequently use integrated circuits. Block and schematic diagrams of such an IC regulator are shown in Fig. 4-17. This circuit can be used by itself or, if more current output is desired, can be used with an additional higher powered series transistor, as illustrated in Fig. 4-18.

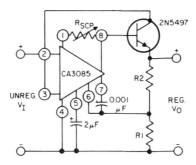

Fig. 4-18 *Integrated-circuit voltage regulator with external high-power transistor (Courtesy, RCA).*

Protective Measures and Devices

The basic nature of a power supply is such that, if something should malfunction either in the supply or in circuits connected to it, damage to components, overheating, and even a fire can result. For this reason, power supplies often include design features and devices to protect the system in case of malfunctions.

One typical malfunction is current overload. A common way to provide protection against overloading is shown in Fig. 4-19. Transistor Q2 samples the voltage across R2, which is proportional to the output current of the supply. Q2, in turn, controls Q1, since its collector is connected to the Q1 base circuit. Q2

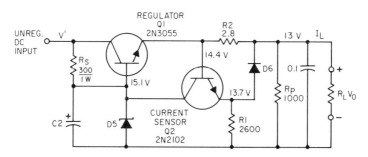

Fig. 4-19 *Current overload protection circuit.*

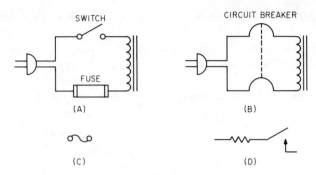

Fig. 4-20   *How fuses and circuitbreakers are incorporated into circuits.*

operates only when the R2 voltage reaches a predetermined value. When that happens and Q2 conducts, it draws current from the Q1 base circuit, causing the resistance of Q1 to increase and limit output current.

Fuses are often contained in power supplies, and sometimes circuit-breakers are used. Figure 4-20 shows how these two devices might appear in typical schematic diagrams of power supplies. A fuse connection is shown in Fig. 4-20(A) and an alternative symbol for the fuse in (C). In Fig. 4-20(B) is the circuitbreaker, with an alternative symbol in (D). Although the ac input circuit is the most common place for fuses and circuitbreakers, they are sometimes found in other parts of the circuit.

Some power supplies are equipped with thermal cutouts, which react to a condition of excessive temperature; when it is too high, it shuts the power off. These breakers are often shown schematically like the circuitbreaker of Fig. 4-20(D). Thermal breakers often have reset buttons; when the equipment has cooled down, the button is depressed and operation restored. An alternative symbol for such a breaker is shown in Fig. 4-21.

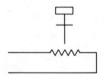

Fig. 4-21   *Symbol for circuitbreaker with reset button.*

Power supplies often incorporate capacitors in the power line input circuit to suppress transients, and to limit the effect of RF currents that might be on the power line. Sometimes a special diode, designed to clip peaks of the power line transients, is used. Examples are shown in Fig. 4-22.

When diodes are connected in series to provide a higher combined overall voltage rating, certain protective measures are often used. Because of slight variations in the characteristics of the rectifiers, the voltage tends not to

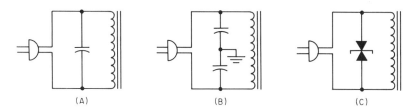

Fig. 4-22   *Examples of line-transient filters.*

divide equally among them. Consequently, some may experience a voltage overload. To minimize this effect, it is common practice to connect a resistor across each rectifier. The resistors all have the same value, so that the voltages are equally distributed during the off period. The value of each resistor is high enough so there is no interference with the desired operation of the rectifiers. Such an arrangement is shown in Fig. 4-23(A). Also shown in this circuit is a capacitor connected across each rectifier; these capacitors filter any transient spikes that appear across the rectifiers.

Voltage distribution resistors are often used across filter capacitors, too. It is sometimes to economic advantage to connect capacitors of relatively low

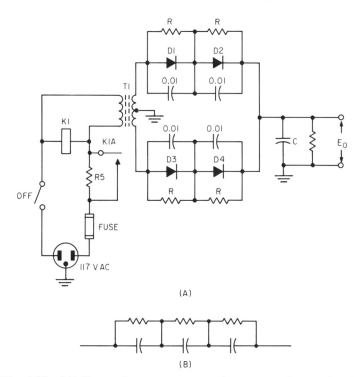

Fig. 4-23   *(A) How resistors are connected across rectifiers to distribute voltage equally, and (B) use of resistors across filter capacitors for the same purpose.*

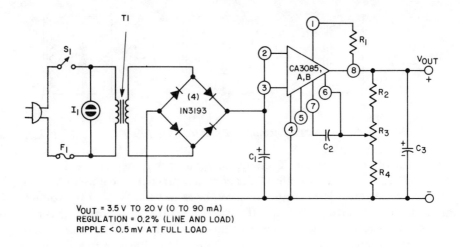

V_{OUT} = 3.5 V TO 20 V (0 TO 90 mA)
REGULATION = 0.2% (LINE AND LOAD)
RIPPLE < 0.5 mV AT FULL LOAD

Parts List

C_1 = 50 μF, electrolytic, 50 V	I_1 = neon lamp, 120 V	R_4 = 1000 ohms, 0.5 watt
C_2 = 100 pF	R_1 = 5.6 ohms, 0.5 watt	T_1 = power transformer,
C_3 = 5 μF, electrolytic, 35 V	R_2 = 8200 ohms, 0.5 watt	Stancor TP-3 or equiv.
F_1 = fuse, 1 ampere, 120 V,	R_3 = potentiometer, 10000	
slow-blow	ohms, 0.5 watt	

Fig. 4-24   *Typical ac-to-dc power supply.*

voltage rating in series to make a combination with a high rating. Since the internal resistance of these capacitors can be different, voltage across some capacitors might become excessive and damage them in the absence of the distribution resistors. Such a connection is illustrated in Fig. 4-23(B).

The circuit diagram of a typical ac-to-dc power supply is shown in Fig. 4-24.

DC-DC Power Supplies

There are many applications in which equipment using direct current must be operated from a dc source. Probably the most common are automobile radio receivers, CB equipment, and similar electronic devices. In most cases, the need for an equipment power supply is eliminated by the use throughout the equipment of transistors that can operate directly on the 12 V potential from the car battery. In other cases, however, it is necessary to have voltages higher (or lower) than 12 V; then, a power supply is required.

In the past, converters using dynamotors or vibrators were popular for this purpose. These are now virtually obsolete. The type of circuit employed in modern equipment is called an *inverter*. Modern inverters use transistor oscillators that do what the vibrator did but do it more efficiently. When oscillating, the transistors act as switches for the current applied to the primary of the transformers. A typical arrangement is shown in Fig. 4-25. Notice the

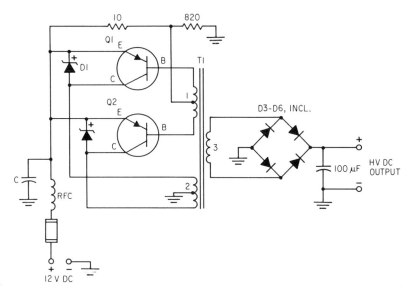

Fig. 4-25 *Dc-to-dc inverter using transistor oscillator switching.*

separate transformer primary winding which feeds energy back from the transistor collectors to the bases to cause the oscillation.

The oscillator circuit is a push–pull type, with one transistor conducting while the other shuts off. The result is a primary current waveform that is approximately square. The current from the battery is thus "chopped" so that a voltage is induced from the primary into the secondary winding of the transformer. In the example of Fig. 4-25, the secondary voltage is applied to a bridge rectifier and single capacitor filter, discussed earlier.

Battery Chargers

Battery chargers are basically a form of power supply. Figure 4-26(A) shows the equivalent circuit of a typical unit. Notice that the charger output potential must not only overcome that of the battery, but must do so with sufficient margin to drive the desired charging current. As indicated by the equivalent circuit and the accompanying equation, the charging rate (current) depends on the difference between the charger voltage and the battery voltage. The battery voltage drops off as the battery discharges, so that the charging rate is much higher for a "dead" or "almost dead" battery than if it is nearly fully charged. The block diagram is shown in Fig. 4-26(B) and the schematic diagram for the simplest form of charger in Fig. 4-26(C).

The schematic diagram of a more elaborate battery charger is shown in Fig. 4-27. In this circuit, a bridge rectifier is used. The two transistors form a sensing circuit to feed current back to the SCR. This limits current to the battery being charged when it is reaching full charge.

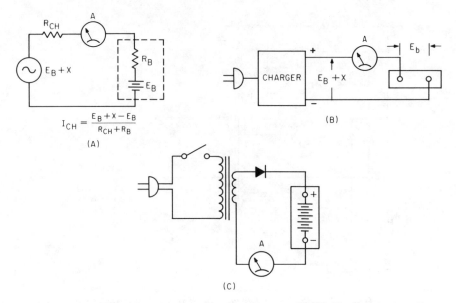

$$I_{CH} = \frac{E_B + X - E_B}{R_{CH} + R_B}$$

(A)

(B)

(C)

Fig. 4-26 *(A) Circuit of a battery charger, (B) block diagram, and (C) schematic for the simplest charger.*

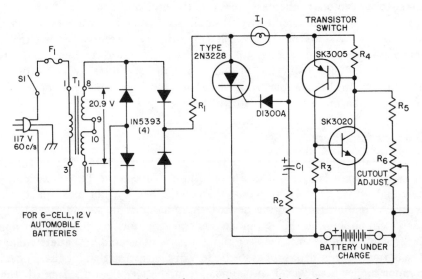

Fig. 4-27 *More elaborate battery charger with a bridge rectifier.*

High-frequency power supplies, such as those operating from the 15,340-Hz horizontal-deflection circuits in TV receivers, are discussed and illustrated in Chap. 8.

CHAPTER

5
Audio
System
Diagrams

Audio amplifiers have evolved from simple one- or two-stage affairs to very sophisticated arrangements in today's electronics. Audio amplifiers were first created for radio receivers, where they were designed to amplify the relatively weak audio signal output of detector stages and make this signal powerful enough to operate a loudspeaker. Transmitters used (and still use) audio amplifiers to increase low-level microphone outputs to the power level needed to modulate the RF output. In amplitude modulation, the AF signal has to be amplified to as high as one half the power input of the transmitter's RF output stage.

Today, the wide demand for high fidelity has added greatly to the sophistication of audio systems. The modern home high-fidelity system is expected to operate in any of a variety of modes, faithfully reproducing programming provided by FM and AM radio, record players, and tape recorders, in either the monaural (mono) or stereophonic (stereo) mode. Provisions are made to match impedances between various pieces of equipment, and to provide loudness and tone controls, filters, muting, and other adjustments.

From this background we approach the information necessary to recognize and interpret the schematic diagrams of typical audio systems. Because of the complexity of the more sophisticated systems, we shall be studying them mainly at the block diagram level and making separate analysis of the individual building blocks. We start, however, with the basic types of AF amplifier stages and auxiliary stages.

There is such a wide variety of AF systems in use that it would not be possible to cover each one individually. This is not necessary anyway, since the principles are the same. In this chapter, therefore, we use a home-type high-fidelity amplifier as our main example, because it contains most of the types of circuits you may encounter.

Types of Audio System Stages

The circuits of an AF system can be better visualized if the system is divided into several types of stages, as illustrated in Fig. 5-1.

For some functions, such as amplifying the output of a tuner or a tape player, the signal input to the amplifier usually is an appreciable fraction of a volt (0.1 V or more). The gain needed in such an amplifier is moderate and can be provided by what might be called a "standard" amplifier. But some devices produce only a relatively weak signal (one or several millivolts), so further amplification by a *preamplifier* is provided, as shown at the left in Fig. 5-1. Following the preamp, the next stage (or stages) in the amplifier is called a *voltage amplifier* to distinguish it from the preamplifier (which also amplifies voltage). The voltage amplifier(s) provides sufficient output signal voltage to operate the first power stage, which, in the case illustrated, is the *driver*.

There may be more than one driver; in such cases, the first driver is often called the *predriver*. The final or *power* stage then provides the desired output power, which, in available audio systems, may be from five to hundreds of watts. Other stages may be included to shape frequency response or control tone; these are usually voltage amplifier stages designed specifically to accomplish these additional functions.

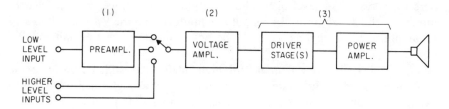

Fig. 5-1   *Diagram showing how preamplifiers, voltage amplifiers, driver stages, and power amplifiers fit into an audio system.*

AF Amplifier Circuit Types

There are three basic AF amplifier circuit types: transformer-coupled, resistance–capacitance-coupled (RC), and (resistance) direct-coupled. Examples of these are shown in Fig. 5-2. The transformer itself is the distinguishing component in the transformer-coupled amplifier illustrated in Fig. 5-2(A). Notice that two PNP transistors are used. The collector current of each transistor passes through the primary winding of a transformer. The audio signal developed across the secondary is coupled into the base circuit of the next stage. Resistors R1 through R6 are used to maintain proper transistor biases.

An RC-coupled circuit is shown in Fig. 5-2(B). Here, NPN transistors are used. Current is fed to the collectors through resistors (R4 and load). Signal input is coupled through C1 to the first stage, whose output is coupled to the

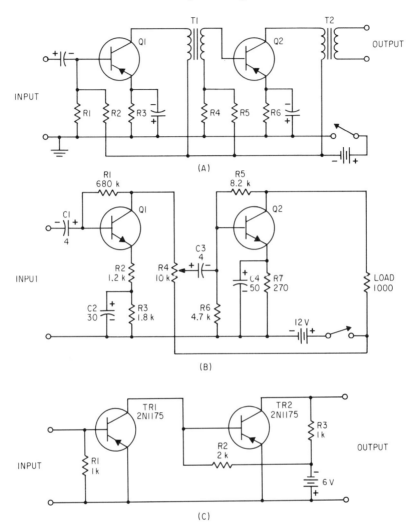

Fig. 5-2 *Basic AF coupling circuits: (A) transformer coupling, (B) RC coupling, and (C) direct coupling.*

second stage through C3. The distinguishing features here, of course, are the coupling capacitors, C1 and C3.

The direct-coupled circuit in Fig. 5-2(C) shows the collector of the first stage connected directly to the base of the next stage. Thus, there is nothing between the first and second stages to limit frequency response. This type of circuit is particularly useful where very-low-frequency signals, not usually passed by a coupling capacitator, must be amplified. Direct coupling can, of course, be spotted in common-emitter stages by the direct connection between the collector of one stage and the base of the next.

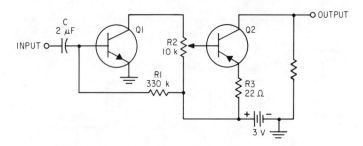

Fig. 5-3   Complementary interstage coupling.

The maintenance of proper biases in a direct-coupled amplifier is simplified in the complementary amplifier illustrated in Fig. 5-3. This circuit uses oppositely biased transistors, in this case an NPN transistor followed by a PNP transistor. This allows direct coupling without special biasing circuits. The emitter-collector circuit of the first transistor is in series with the base-emitter circuit of the second transistor. Such an arrangement naturally provides and maintains the required biases. This circuit is easily identified by (1) the fact that the transistors are different (one is a PNP and the other an NPN) and (2) that normally the coupling is direct.

Layout of a High-Fidelity Audio System

Before proceeding to specific examples of audio system schematic diagrams, it is appropriate to consider the general plan for an audio system, so that its parts can be recognized in schematic analysis.

Block diagrams of audio systems from simple to more elaborate are shown in Fig. 5-4. In (A) we see a very simple system in which the input signal is a high enough level so that one stage of amplification can provide the output needed to drive the power amplifier. This block diagram is typical of the amplifiers used in some simple record players (with a high-output pickup cartridge) and of small amplifiers that can provide speaker output for a radio tuner. In a more common arrangement, illustrated in Fig. 5-4(B), a preamplifier is added to allow use with low-level devices such as low-level phono pickups and microphones. Also added is a driver section, desirable for operating a higher level output stage.

Each block in Fig. 5-4 may be considered to represent a single stage, or two or more stages of amplification. As we shall see, several transistors may be used for each section, with more than one transistor often used for a single stage.

In Fig. 5-4(C), the system is expanded to accommodate stereo operation. This simply means duplicating the single amplifier in Fig. 5-4(B) to provide two of these amplifiers. One becomes the left channel and the other the right channel. Of course, in the case of the stereo phono pickup indicated, only one source device is used, but it has two outputs. One is the input signal to each amplifier.

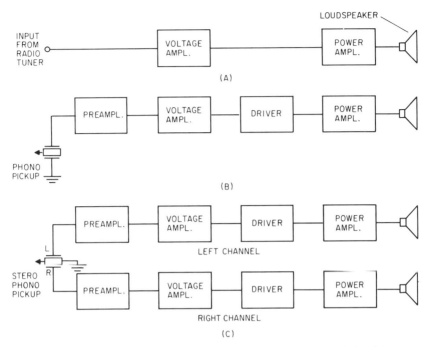

Fig. 5-4 *Block diagrams of audio systems: (A) simplest, high-level-input monaural, (B) the same, with preamplifier, and (C) stereo system.*

Figure 5-5 shows the layout of one channel of a typical amplifier for a high-fidelity system. An important feature is the variety of inputs and how they are handled. The left (L) channel is shown in block form; in a complete amplifier another identical channel is used, as indicated in this diagram. Starting in the upper left corner of Fig. 5-5, note the phono inputs. The L channel input is connected to the L channel phono preamplifier whose output goes to the function switch. The right (R) phono input goes to an identical preamplifier in the R channel. The preamplifier is needed because most high-fidelity phono cartridges put out only several millivolts of signal, which must be amplified considerably to provide the necessary input signal to the voltage amplifiers. The phono preamplifier also includes frequency response correction. This correction is needed because, to avoid overloading record grooves at low frequencies (or insufficient power at high frequencies), recordings are made with a special characteristic (RIAA) and the preamplifier must correct for this.

Next are the radio inputs. The FM input has two, the L and R inputs for stereo, and the AM input, which is monaural. Notice that the input terminals on the AM part of the switch are connected together, so that the single AM signal is applied to both channels. A separate preamplifier is used for the radio input.

The auxiliary (AUX) input is for high-level signals, such as the output of radio or tape player. Because this is for input levels from 0.25 to 1.0 V, no preamplifier is necessary.

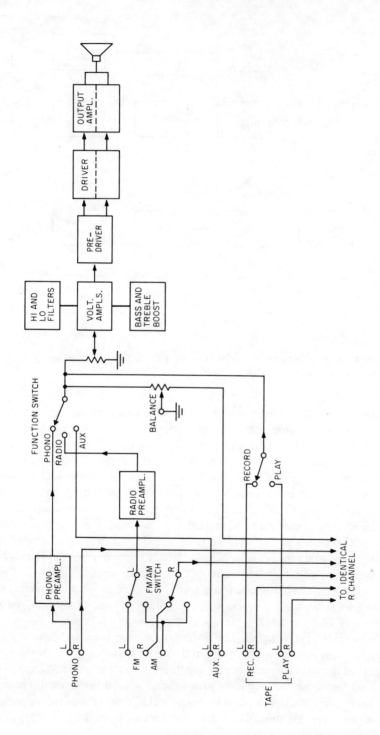

Fig. 5-5 Layout of one channel of a typical hi-fi system amplifier.

The tape inputs are next. The *tape play* input feeds the signal from a tape deck output into the voltage amplifier so that the tape may be played through the system. The *tape record* terminals allow signals available at the other inputs to be recorded on an external recorder. These connect the recorder input to the input of the voltage amplifier, so that signals arriving there from the phono, FM, and AUX inputs can be recorded.

The voltage amplifier section usually consists of two or more stages. Contained in them or connected to them are the tone controls and any filters that might be included. High-fidelity amplifiers usually have independent bass and treble boost controls. High- and low-frequency filters are often included to allow cutting response at the higher or lower frequencies or both. These filters and controls allow the listener to adjust response to "taste." They should not be confused with preamplifier correction type networks (such as the phono RIAA correction) which are used to correct a known and deliberately introduced input signal frequency distortion.

The volume control for the system is also usually associated with the voltage amplifier circuits. The volume controls for the two channels are ordinarily ganged to provide simultaneous control of both channels.

Also in the area of the voltage amplifiers is the balance control. This allows the user to adjust for proper balance between the output levels of the two loudspeakers.

From the voltage amplifiers, the power of the signal must be built up sufficiently to drive the power amplifier. Since the power amplifier drive requirement is appreciable, both predriver and driver stages are often required, as shown in Fig. 5-5.

Two outputs are shown from the predriver to represent double-ended or push–pull output. In this circuit (or sometimes in the driver or voltage amplifier) there is a phase inverter circuit. This circuit takes a single-ended signal (single voltage against ground) and converts it to a balanced (two voltages of opposite phase balanced against ground) signal such as is used in a push–pull circuit. The driver and power amplifier are then push–pull stages.

All these stages are repeated identically in the right channel and each channel has its own loudspeaker system to provide stereo output.

Preamplifiers

Preamplifiers are used in audio systems to amplify low-level signals from such devices as microphones and low-output phono pickups. They must bring these signals up to levels comparable to those from radio tuners, so they are adequate for the amplifier stages that follow. Preamps are designed also to operate at low current levels and to maintain a low noise level, since these stages set the noise figure for the whole amplifier.

A preamplifier may have one stage or several stages. Because there is nothing in the design that makes the circuit look much different from those of

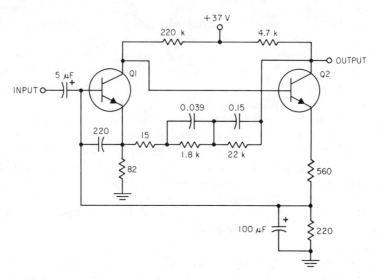

Fig. 5-6   *Example of a preamplifier for a phono input in a hi-fi system.*

other sections, preamplifiers must be identified at least partly by their location in the overall system. They are the first amplifiers the input signal encounters, so they will normally be found on the left side of the schematic. They are associated with the connections from the input source devices, such as phono pickups, radios, microphones, and tape recorders.

An example of a preamplifier circuit used for a phono input is given in Fig. 5-6. Two transistors in common-emitter configuration are connected in cascade. Notice that there is a feedback connection from the collector of Q2 to the emitter of Q1. In the feedback path are resistors and capacitors, connected to provide compensation for the output characteristic of the phono pickup. Direct coupling is used, thus maintaining good low-frequency response.

Voltage Amplifiers

Preamplifiers, and, to a lesser extent, drivers and many power amplifiers, are voltage amplifiers, because the voltage output in each case is greater than that of the input. (Power amplifiers are designed mainly to increase power rather than voltage, so they usually have either low voltage gains or none at all. However, we are talking here about those stages between the preamplifiers and the power stages.) Preamplifiers have the prime purpose of amplifying low-level inputs (for example, those of only a few millivolts) up to a level of 100 mV or more, sufficient to drive the voltage amplifier stages, with a minimum noise contribution; also, a preamp provides any fixed frequency-response shaping such as that needed for RIAA compensation.

Voltage amplifiers must accept the preamplifier output and bring its level up sufficiently to drive the power stages, which usually require from one to several volts of signal.

Voltage amplifiers are often common-emitter RC stages like those already discussed. Occasionally, special circuits are used. An example is given in Fig. 5-7. This is a pair of transistors in the common-collector configuration. This arrangement is known as *Darlington pair*. Such pairs are available in integrated circuits, which are popular for this purpose.

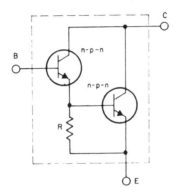

Fig. 5-7   *Darlington pair voltage amplifier (Courtesy,* RCA).

Another voltage amplifier arrangement might be encountered: the *differential amplifier*. A typical circuit is shown in Fig. 5-8. Figure 5-8(A) shows how such an amplifier is packaged in IC form; Fig. 5-8(B) is the schematic diagram of the differential amplifier circuit, and Fig. 5-8(C) the IC package in a typical circuit. In this type of circuit, a third transistor provides a constant-current source for the emitters to stabilize operation. The collector of the first transistor is coupled to the next stage (usually the driver) and the collector of the second stage is returned through a load resistor to the power supply voltage. This arrangement offers good stability and low distortion.

Tone Controls

High-fidelity amplifiers virtually always include bass-boost and treble-boost circuits. These may be located in a preamplifier, but more likely in the voltage amplifier portion. They are usually composed of RC networks, one element of which is a variable resistance with which to adjust the degree of boost provided.

An example of bass- and treble-boost circuits is shown in Fig. 5-9. It is part of the input to a Darlington pair voltage amplifier.

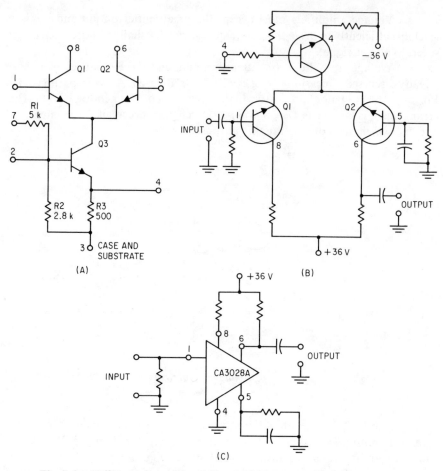

Fig. 5-8   *Differential amplifier: (A) typical IC arrangement, (B) circuit for AF voltage amplifier, and (C) IC with necessary external connections.*

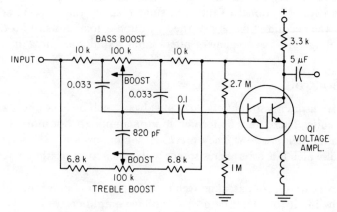

Fig. 5-9   *Bass- and treble-boost circuits.*

Phase Inverters

As has become evident, practically all preamplifier and voltage amplifier circuits are single-ended. That is, they are designed to amplify a single voltage with respect to ground. All amplifiers of any size or sophistication, however, use power stages (driver and output stages which are double-ended, or push–pull). In these amplifiers, the signal is split into two parts: one 180° out of phase with the other.

This means that at some point the single-ended signal must be converted into one in push–pull. Some voltage amplifier or predriver stages are so designed that the *phase inversion* or *phase splitting* is done as a natural part of the circuit. In other cases, circuit elements are added to provide the desired extra signal phase.

The circuits of some of the more common phase inverters are shown in Fig. 5-10. One of the simplest (but not least expensive) phase inverters is a transformer. The primary winding is grounded (to signal) at one end to form a single-ended circuit. The secondary winding has a center tap which is grounded, so that each half of the winding produces a signal 180° out of phase with the

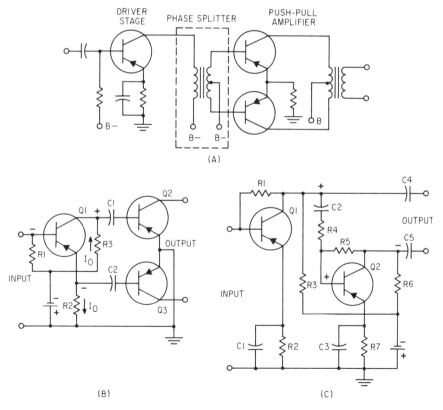

Fig. 5-10 *Phase-inverter circuits: (A) transformer-coupled, (B) single-ended, and (C) two-stage.*

other half. Thus, the transformer provides the proper coupling from a single-ended amplifier to one that is push–pull. A simple version of such an arrangement is shown in Fig. 5-10(A).

A transformer is not necessary to provide phase inversion. A single-ended stage may be made into an inverter by taking one output from the collector and the other from the emitter [Fig. 5-10(B)].

Another way to provide the inversion is through the use of two transistors, as illustrated in Fig. 5-10(C). The first transistor provides output from its collector; since this is a common-emitter circuit, the collector output is out of phase with the input (base) signal. Besides being used as one of the double-ended outputs, the output of the first transistor is also coupled to the base of the second transistor. The second stage, also connected as a common-emitter amplifier, again inverts the signal, to produce the other desired output.

There are a number of ways in which the desired phase inversion can be provided. Another is through the use of the differential amplifier whose circuit is shown in Fig. 5-8. The reader may encounter others, but the principles are always the same.

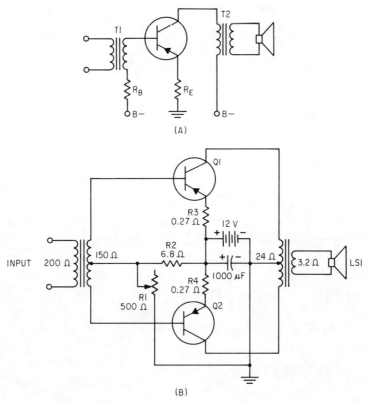

Fig. 5-11 *Transformer-coupled AF power output circuits: (A) single-ended, and (B) conventional push–pull.*

Drivers and Power Amplifiers

There are several general types of amplifiers used in the driver and power output stages of AF amplifiers. The simplest is the single-ended transformer-coupled type illustrated in Fig. 5-11(A). This type is used where limited amounts of power are needed, such as for small, low-cost record players. It is also commonly found in small radio receivers.

In Fig. 5-11(B) is the circuit of a transformer-coupled push–pull amplifier. Transformers are used both for input and output. It is important to notice that the input transformer is a phase inverter to provide push–pull signal voltages for the bases of Q1 and Q2 from its single-ended input winding. Also notice that the output transformer changes the push-pull output signal configuration into a single-ended one needed for the loudspeaker system.

Because output transformers are relatively costly, most of today's audio systems use push-pull output circuits that eliminate them. This is done by connecting the two transistors in series as far as the direct current supply is concerned, and using the connection between the emitters as the output.

Two examples of series-connected output circuits are shown in Fig. 5-12. In (A), two NPN transistors are used, connected so that electron current

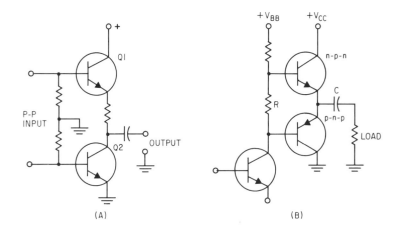

Fig. 5-12 *Series push–pull arrangements: (A) conventional, and (B) complementary (Courtesy,* RCA).

follows the path from ground into the emitter of Q2, through Q2 to the emitter of Q1, then through Q1. The output to the loudspeaker system is coupled from the emitters through a capacitor, thus eliminating the need for an output transformer.

Another series arrangement takes advantage of the opposite polarities of the NPN and PNP transistor types. In this arrangement, illustrated in Fig. 5-12(B), the drive is supplied in parallel from a single-ended driver. The

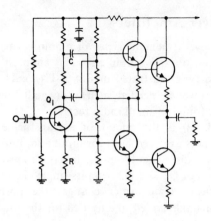

Fig. 5-13   *Output amplifier using Darlington pairs* (*Courtesy,* RCA).

need for a push-pull driver or separate phase inverter is thus eliminated, and there is no need for either input or output transformer.

A popular arrangement is that illustrated in Fig. 5-13. In this circuit the driver and output transistors are connected as Darlington pairs. Notice also the phase inverter, which is like that in Fig. 5-10(B).

6
Radio
Receivers

In Chap. 3, the integration of components into radio receiver circuit sections or modules is considered. Since this chapter is solely devoted to receivers, let us recap some general information about them.

One of the important factors in recognizing and interpreting a circuit diagram is a familiarity with certain general characteristics and necessary features of a radio receiver. The purpose of a radio receiver is to accept and process very minute signal currents from an antenna and make the intelligence in these signals available as sound waves from a loudspeaker or headphones. (We are discussing here only radio receivers, as opposed to television, facsimile, and control-signal receivers discussed elsewhere.) Included in this category are AM, FM, shortwave, and communications radio receivers.

Keeping in mind the nature of its input and output signals, the diagram of a radio receiver is identified by symbols for an antenna at the input (left) and a loudspeaker or headphones at the output (right). The most common symbols used for external antennas (those not built into the receiver) are indicated in Fig. 6-1. Symbols D and E are most common for cases in which a very high frequency (VHF), ultrahigh frequency (UHF), or similar antenna is to be used,

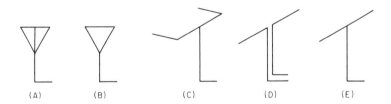

(A)   (B)   (C)   (D)   (E)

Fig. 6-1   *Schematic symbols for external antennas.*

although the symbols of (A) and (B) are also sometimes used in these cases. These symbols are used with either a schematic or block diagram.

In many schematic diagrams, an antenna symbol is not used, because the diagram ends where leads to the antenna leave the set. For example, outdoor antennas for FM receivers ordinarily are connected by 300-ohm "ribbon" transmission line to two terminals on the rear of the receiver. In this case, the diagram label identifies the antenna connection point.

As an aid in learning to analyze a receiver, consider the block diagrams of Fig. 6-2. The diagram in Fig. 6-2(B) is a repeat of that shown in Fig. 3-5. It will guide us as we consider typical generalized circuits for each section of a receiver.

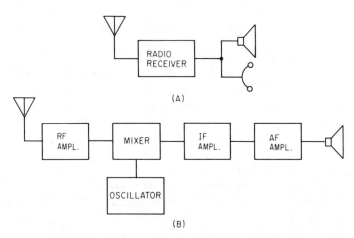

Fig. 6-2   *Basic receivers in block diagram form.*

The RF Amplifier Section

A general knowledge of basic circuits making up the essential parts of broadcast receivers helps one to know what to look for in identifying them. In this chapter we use simplified circuits to illustrate essentials, without dwelling on any particular versions. Some variations are apparent in the typical overall circuits to follow.

As shown in the block diagram of Fig. 6-2(B), the first section in the functional signal flow is the RF amplifier. Not every receiver has an RF amplifier; small AM receivers often omit it. Some more elaborate receivers may have two or more stages of RF amplification.

A typical RF amplifier stage diagram is shown in Fig. 6-3(A). Amplifier transistor Q1 has a tuned input coil L1 and a tuned mixer coil L2, into which the collector lead from Q1 is tapped. The antenna is coupled to a tap on L1. This is an appropriate connection for cases in which a single-wire antenna lead is used. This arrangement is most likely in shortwave receivers, which are designed for a single-wire or a whip antenna. Other input circuits used with outside antennas are shown in Fig. 6-3(B). The one at the left is for a single-wire antenna and that at the right for an antenna with a coaxial cable lead-in to the receiver.

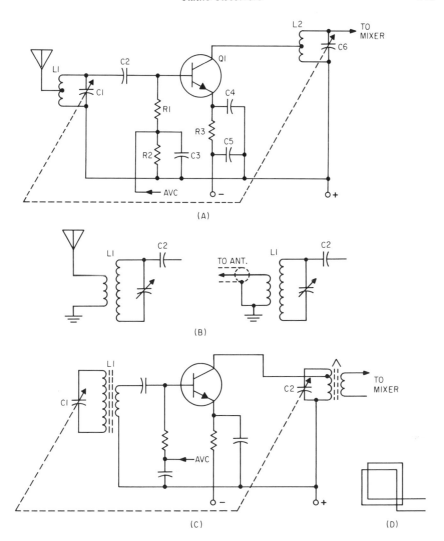

Fig. 6-3 *Schematic diagrams of RF amplifiers: (A) with external antenna, (B) other antenna coupling arrangements, and (C) with internal loop-type antenna; (D) loop antenna symbol.*

Returning to Fig. 6-3(A), notice that bias is maintained on the base of Q1 by the base current through R1 and R2. Automatic volume control (AVC), derived from the detector, is applied across R2, so the bias on the base varies appropriately with the strength of the signal. Emitter bias is maintained by R3 in this common-emitter circuit. The dashed lines indicate that C1 and C6 are ganged so that the tuned circuits are maintained at the same (received-signal) frequency. The capacitor tuning gang usually also includes a section which tunes the oscillator (not shown here) to its appropriate frequency. This RF amplifier circuit is identified by two main features: (1) its location (normally) at the extreme left of the receiver diagram, and (2) the symbol for the antenna connected at the input. Figure 6-3(B) shows other antenna coupling arrangements.

Another version of RF amplifier is shown in Fig. 6-3(C). This is very much like that in Fig. 6-3(A), except that no antenna connection is indicated. This circuit is typical of those used in most AM receivers, in which the antenna is self-contained. The RF transformer and tuning coil L1 form the antenna. It is ordinarily a coil wound on a ferrite core, sometimes called a *loopstick*, to give it a high Q. Usually, an antenna of this type is of sufficient physical size to intercept signals of usable strength. Many earlier models of small AM receivers used a *loop* antenna. This was a flat coil, wound on a piece of cardboard or plastic, and mounted against the back of the receiver. The symbol for such an antenna is shown in Fig. 6-3(D).

Capacitor C1, which tunes this coil antenna, is one section of the ganged capacitor which also tunes the mixer input circuit and the oscillator. Practically all AM radio sets now use this arrangement. Although the antenna symbol is not used for this circuit, the indication of an iron core for the input coil, plus the extreme-left position of the circuit section in the receiver diagram, tell us that this is an RF amplifier. Another way in which this circuit differs from that of Fig. 6-3(A) is that here the output circuit is a transformer.

The Mixer/Converter Section

Next, after the RF amplifier in the functional flow, is the mixer. This section may be separate or it may be a part of a single-transistor converter circuit. For the mixer or converter section to operate, it needs an oscillator to supply a signal to mix with the received signal to produce an IF signal. In other words, a converter is a mixer and an oscillator combined. Usually, when the oscillator and mixer functions are both accomplished by one transistor, the combination is called a converter, which has functional parts called the mixer and the oscillator. When separate solid-state devices (the mixer may be a diode or a transistor) are used for mixer and oscillator, these names predominate.

The case of a separate mixer and oscillator is illustrated by the circuit of Fig. 6-4. The mixer circuit (Q1) is just like that of a simple transistor RF amplifier, whose input is the output signal from the RF amplifier of the receiver. The signal from oscillator Q2 is coupled through a capacitor to the base of Q1. In other words, we recognize a separate mixer as simply a common-emitter amplifier arrangement with both received and oscillator signals coupled to the base. Of course, then the collector is connected to the first IF transformer. In some special types of receivers, other mixer arrangements are used; these are discussed later.

As previously mentioned, the oscillator and mixer functions can be combined, and this is the more common arrangement. An example is the circuit

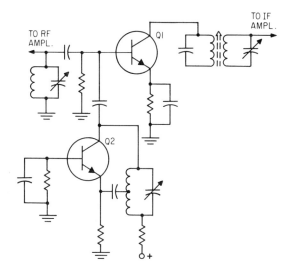

Fig. 6-4 *Converter with separate oscillator and mixer.*

of Fig. 6-5. One way to spot this setup is through its "extra" coil. An RF amplifier or simple mixer would have coils or transformers at the input and output; this circuit has them (T1 and T2) as well as oscillator coils L1 and L2. Coil L1 couples collector (output) current variations back through L2 to the base, causing oscillation at a frequency determined primarily by L2-C3. At the same time, the signal from the RF amplifier is amplified and mixed in Q1 with the oscillator signal. Since the primary of T2 is in series with the collector circuit, the IF signal is coupled through T2. Other types of oscillator circuits are shown in Chapter 2.

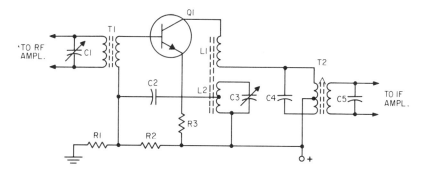

Fig. 6-5 *Mixer and oscillator functions combined.*

To summarize, the converter section can be recognized by the following observations:

1. There must be an oscillator circuit, either separate or combined with the mixer;
2. The section is preceded by either an RF amplifier or the antenna;
3. It is followed by the IF amplifier.

IF Amplifier Section

The average IF amplifier is normally of from two to four stages. If the receiver is of the simple AM variety (without FM or other bands), the IF amplifier stage diagrams may be identical. Otherwise, the IF stages may differ from each other.

The most common circuits may use any of three different arrangements, illustrated by the block diagrams of Fig. 6-6. In simple AM broadcast type receivers, there are two or more AM IF amplifiers, operating on 455 kHz, followed by an AM detector. This is illustrated in Fig. 6-6(A).

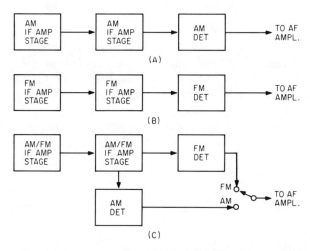

Fig. 6-6 *IF amplifier arrangements: (A) for AM, (B) FM, and (C) AM/FM receivers.*

The exclusively FM receiver IF section is laid out in the same way in block diagram form as shown in Fig. 6-6(B). Although it appears identical here to the AM variety, it differs considerably in construction of parts, due to the fact that it operates at the FM intermediate frequency of 10.7 MHz instead of 455 kHz. There are some circuit changes also, and there are often more stages than in AM receivers, because individual stage gain is considerably less in FM stages.

Of the small home-type broadcast receivers, probably the most common are the AM/FM type. In these, there are separate AM and FM sections for the RF amplifier and converter. (These arrangements are discussed later in this chapter.) But the IF amplifier section of such a receiver usually employs

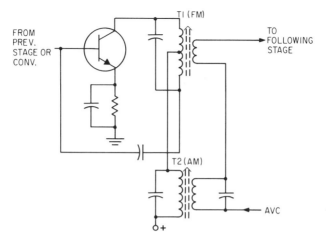

Fig. 6-7 *Schematic diagram of an AM/FM IF stage circuit.*

combination AM/FM IF amplifier stages and each stage can operate on either 455 kHz or 10.7 MHz. Figure 6-6(C) is a block diagram illustrating such an arrangement.

The typical IF amplifier schematic diagram is about the same as that for AM [see Fig. 2-40(B)]. We shall, therefore, discuss here only the diagram for a combination AM / FM IF amplifier stage. Such a stage is illustrated in Fig. 6-7. Notice that two IF transformers are used: T1 for FM and T2 for AM. The two primary windings are connected in series, as are the two secondary windings. No switching is necessary to change between AM and FM operation. When the signal being amplified is FM (10.7 MHz), the tuning capacitor for the primary of T2 has a very low impedance, thus effectively eliminating (shorting) the AM coil and capacitor from the circuit. When the signal is AM (455 kHz), the coil of the FM tuned circuit has, relatively, a near-zero impedance, shunting the FM circuit out.

In some cases, ceramic filters are used between IF stages instead of transformers. In such a case, the filters are usually indicated as boxes and labeled as such. An example of a diagram of an IF stage using a ceramic filter is shown in Fig. 6-8. Advantages of ceramic filters are low cost and the greater selectivity possible.

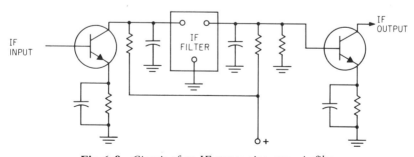

Fig. 6-8 *Circuit of an IF stage using ceramic filter.*

Detectors

The next function is that of detection (demodulation) in which the intelligence is removed from the modulated IF signal. Both AM and FM detectors utilize diodes as the basic elements, but the methods are different.

The demodulation is accomplished for AM by passing the modulated signal through a diode, then through a capacitor-bypassed load resistance. The demodulated signal (audio, video, or pulses) then appears across the resistance.

A typical AM detector circuit is shown in Fig. 6-9. The primary of transformer T1 is connected to the last IF stage. As indicated, the secondary of the transformer is connected across the combination of D1 and load resistors R1 and R2. Resistor R2 is variable and acts as the volume control. The voltage drop across R1 and R2, filtered by R3 and C3, becomes the automatic volume control (AVC) voltage. The latter is fed back to RF, mixer, and IF amplifier bases, to reduce gain for high-level signals and thus prevent overloading.

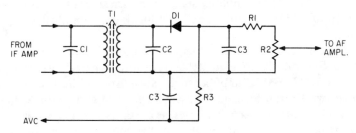

Fig. 6-9   *AM detector circuit.*

In FM receivers, two types of detectors are in general use: the *ratio detector*, and the *Foster-Seeley detector*. These circuits are shown in Fig. 6-10. The ratio detector in Fig. 6-10(A) uses two diodes connected in series to the secondary of the last IF transformer, with load resistance R1 + R2, also in series and bypassed by C4 and C5, respectively. The center tap of the T1 secondary is connected through C2 to the top of the primary. C1 and C3 tune the transformer windings to resonance at the FM "rest" frequency. In a ratio detector circuit, as the frequency of the received signal shifts in one direction, the voltage increases across R1 and decreases across R2; a frequency shift in the other direction has the opposite effect. Thus, the dc voltage at the midpoint between R1 and R2 (or C4 and C5) shifts instantaneously in accord with the instantaneous frequency of the received IF signal. Since the received signal frequency varies in accord with the AF modulation, this means the AF signal is available at that point.

An important feature of the ratio detector is the fact that it is self-limiting. Carrier amplitude limiting is accomplished by C6, which is an electrolytic capacitor of relatively large capacitance (about 5 μF). The sum of the voltages across C4 and C5 is constant, even though the ratio of these voltages changes with modulation. That sum voltage is maintained by the large charge

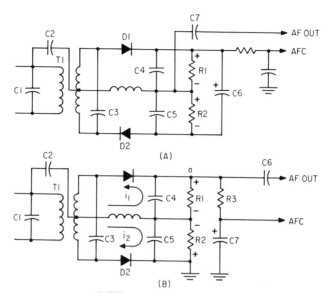

Fig. 6-10   FM detectors: (A) ratio, and (B) Foster-Seeley.

on C6. Sudden amplitude changes in the signal are absorbed by this electrolytic capacitor.

The Foster-Seeley discriminator diagram is shown in Fig. 6-10(B). It appears almost identical to that of the ratio detector, but there are important differences. First, the diodes, instead of being connected in series, are connected back-to-back. The rectified current through R1 (i_1) is opposite to that through R2 (i_2). Thus, the R1-C4 rectified voltage polarity is opposite to that across R2-C5, with respect to the total R1 + R2 voltage. When the received FM signal is at rest frequency, the R1 voltage exactly equals the R2 voltage, so that the voltage between ground and point *a* is zero. A swing of the received signal frequency causes the voltage across one resistor to go up and the other to go down, producing a net voltage at *a*. The result is a demodulated AF signal at *a*. Thus, another difference from the ratio detector is that AF output is taken from the sum of the two resistors, rather than their center common point. Also, there is no large capacitor, such as C6 in Fig. 6-10(A), so that this detector cannot provide its own limiting.

Because, at first glance, the ratio detector and the Foster-Seeley detector appear so much alike, let us review differences that help in identifying which is used in a given case.

1. **Connection of diodes:** In the ratio detector, they are connected *series-aiding*, while in the Foster-Seeley detector they are *series-opposing*.

2. **The connection for audio output:** Audio signal appears *between* the

load resistors for the ratio detector, and *across* both resistors for the Foster-Seeley detector.

3. A large *limiting capacitor* is used with the ratio detector but not with the Foster-Seeley detector.

AF Amplifier Section

The last section in the line of functional flow in a radio receiver is the AF amplifier portion. The AF signal from the detector is at a moderately low level (say about 1 V). It must be amplified to a power level sufficient to operate the loudspeaker. This normally involves two stages: a voltage amplifier and a power amplifier. In some cases, both functions may be accomplished by one transistor stage; in others, there may be one or two voltage and power stages. A typical AF amplifier for a radio receiver is shown in Fig. 6-11. There are two voltage amplifiers, capacitance-coupled. The second voltage amplifier is transformer-coupled to the push–pull power amplifier, which is, in turn, transformer-coupled to the loudspeaker. There are a number of variations of this arrangement in radio receivers. Such variations are discussed in Chap. 5.

Communications Receivers

As the name implies, these receivers are designed primarily for communications, that is, two-way exchange of information as opposed to entertainment. Most common examples of such receivers are those used for citizens band (CB) and amateur radio use. Other examples include equipment for police and fire departments, forest and park rangers, and the like.

Most communication work today is done with a *transceiver*, that is, a combination transmitter and receiver in a single package. Some transceiver sections can be used for both transmitting and receiving, so that the combination provides some economy and compactness compared to separate transmitters and receivers.

For the sake of an orderly approach we shall consider, first, separate communication receivers and transmitters, then discuss how receiver and transmitter circuits can be combined into single transceiver units.

As in the case of AM and FM entertainment-type receivers, communications receivers use the superheterodyne type of circuit. However, communications receivers (particularly the radio amateur type) are designed to get the highest possible performance for minimum received signal levels, so that usable understandability for even very weak signals can be achieved. This feature contrasts sharply with the usual entertainment types, which normally operate with far more than minimum signal levels.

The communications receiver is also likely to have a design more intensely aimed at better sensitivity (low noise figure), selectivity, dynamic range, and stability. In recognizing the schematic diagram of this type of receiver, the above design aims are likely to appear evident in the following ways.

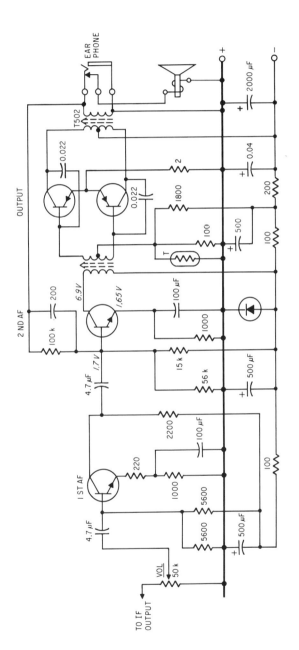

Fig. 6-11 Typical AF amplifier section.

Sensitivity: Most communications receivers contain an RF amplifier, and this is often "trimmed-up" on each received signal using a separate knob or dial on the receiver panel. The RF amplifier is often referred to as the *pre-selector.* The RF amplifier transistor and its circuit are usually selected to optimize signal-to-noise ratio.

Selectivity: Attention is given to two kinds: wide-range selectivity and skirt selectivity. The former is provided by the RF amplifier, which rejects image signals and other spurious responses the IF section can't reject. The skirt selectivity is provided by the IF amplifier; this is the *adjacent-channel* type of selectivity that separates signals close together in frequency. This selectivity is usually provided by such features as crystal filters or mechanical filters in the IF section.

Dynamic range: This is the ability of the receiver to handle signals of widely different signal strengths without loss of sensitivity for the weak ones and without overload for the strong ones. This is accomplished primarily by an AVC system usually more efficient than those in entertainment receivers. The AVC in most communications receivers can be turned off under certain receiving conditions.

Stability: This becomes especially important because of the sharp selectivity sometimes employed. If the receiver is not quite stable enough, the oscillator can drift so that the received signal moves right out of the receiver's pass band.

Modes of Operation of Communications Receivers

Communications receivers of the amateur radio variety in particular are usually equipped to receive any of several types of signals. High-frequency or HF (3–30 MHz), receivers usually are equipped to receive either continuous wave (CW or code) or SSB (single-sideband voice) signals, and many can also receive AM signals. VHF (30–300 MHz) and UHF (300–3000 MHz) receivers are most often designed for reception of FM only, although some amateur receivers in this range can also receive AM, CW, or single sideband (SSB) signals.

We now consider what is required to receive signals in the different modes and what we would look for in the receiver schematic for each.

Single-sideband Receivers: Filters

An SSB receiver must be able to accept a signal with no carrier and with only one sideband. Since there is no carrier, the RF signal varies from zero amplitude (when there is no speech) to maximum amplitude, in constant short-term variations. The dynamic range capability, therefore, must be good.

The most evident features distinguishing an SSB receiver are selectivity and detector design. To optimize signal-to-noise ratio and minimize the bandwidth occupied, the selectivity must be sharp and precise. Receivers for SSB reception normally have a pass band in the range of 2.0–4 kHz. The carrier side

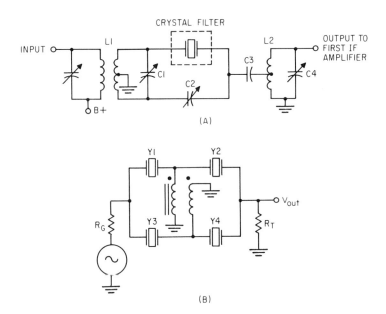

Fig. 6-12   *Crystal filter circuits: (A) simple, and (B) lattice.*

of the received sideband must have good attenuation, to minimize interference, and to allow conservation of valuable radio frequency spectrum space.

The method used to achieve this selectivity is either a *crystal filter* or a *mechanical filter.* The filter is connected right after the mixer in a single-conversion receiver or right after the second mixer in a dual-conversion receiver (discussed below). The filter circuit is such that it can be switched in or out of the circuit, so that, when SSB is not used, the greater bandwidth of the remainder of the IF section can be realized for such purposes as receiving broader band AM signals. There is often another filter circuit, with a much narrower bandwidth (50 to 1000 Hz) for use when receiving CW signals.

The schematic diagrams of typical crystal filter circuits are shown in Fig. 6-12. The simplest form is that in Fig. 6-12(A), which uses one crystal. The IF signal is fed through the crystal, which, at its series-resonance frequency (same as the receiver's intermediate frequency), passes signal, but at other frequencies offers a very high impedance. The secondary winding (L1) on the IF transformer is center-tapped and develops oppositely phased signals at its ends. The main signal goes through the crystal and the opposing signal through C2. C2 is adjusted to be equal to the parallel capacitance of the crystal, thus canceling its parallel resonance effect. A more elaborate circuit, known as a *cascade half lattice filter,* is shown in Fig. 6-12(B). The resonance characteristics of the crystals are combined to provide a sharper, narrower pass band.

Another way of obtaining the narrow, steep-skirted pass band needed for SSB is the use of *mechanical filters.* In these, metal discs resonate mechanically at the intermediate frequency. IF signal is fed to these discs through one transducer and delivered from their output through another. On schematic diagrams, such a filter is usually shown simply as a rectangle. An example of a section of a circuit using one is shown in Fig. 6-13(A). Another symbol sometimes used to indicate this type of filter is shown in Fig. 6-13(B).

Single-sideband Receivers: Detectors

The other major requirement in an SSB receiver is the nature of the detector. As mentioned earlier, since the SSB signal ideally has no carrier,[1] the carrier must be restored at the receiver. This is done at the final detector, where the IF signal is demodulated. A diode detector, such as those previously discussed, can be used with a beat frequency oscillator (BFO) to add the carrier, or an active (transistor) detector can be used with the BFO. An arrangement like this is referred to as a *product detector*.

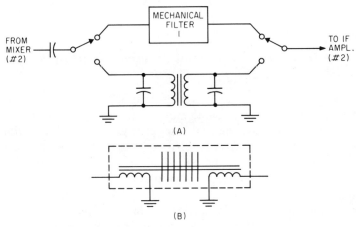

Fig. 6-13 *(A) Mechanical filter connected between mixer and IF amplifier, and (B) alternative symbol for the filter.*

Examples of diode and transistor type product detector circuits are shown in Fig. 6-14(A) and (B), respectively. It will be noted that product detectors are simply nonproduct detectors with BFO injection added. This injection may be by inductive coupling as in Fig. 6-14(A) or by capacitive coupling as in (B). The product detector is really a mixer working just like the mixer in the superheterodyne receiver, except that the latter mixes a signal from the local oscillator with the received signal to produce a signal of intermediate frequency. The product detector mixes the IF signal with a signal very close to the same frequency to produce AF output. Such a detector can, in fact, be used to

---

[1]Since nothing is perfect, there is, in practice, some carrier. Usually, it is kept 40 dB or more below the average signal level. In a few cases of commercial use, some carrier is purposely kept for use as a "pilot" in reception, but the level is quite low.

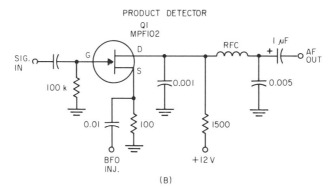

Fig. 6-14 *Product detectors: (A) with diodes, and (B) with FET.*

demodulate received signals directly in what is known as a *direct conversion* or *autodyne* receiver. Such a receiver is like a superheterodyne receiver with the IF section removed and the local oscillator operated on approximately the same frequency as the received signal.

Single-Sideband Receivers: Sideband Selection

SSB transmissions may be on either the upper (frequencies above that of the suppressed carrier) or lower (frequencies below it) sideband. Since the upper sideband and lower sideband are of the same width, reception through the IF section is the same for either. A spectrum from 2 to 4 kHz wide, without a carrier, is selected and amplified. In the product detector, however, the BFO signal frequency must be at one edge of the received sideband; which edge depends on which sideband. Since the BFO replaces the missing carrier, its frequency must be on the lower edge of the upper sideband and higher edge of the lower sideband. (Actually, since the local oscillator frequency in most receivers is higher than that of the received signal, this relationship is reversed in the IF section and product detector.)

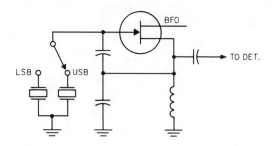

Fig. 6-15   *Switching the BFO frequency allows the product detector to operate on either the upper or lower sideband.*

Thus, to provide quick selection of upper or lower sideband, the BFO is designed to have two switch-selected frequencies: one at the upper and one at the lower edge of the IF pass band. Because of the stability and precision requirements for SSB, it is common to use a crystal oscillator with two selectable crystals. An example of such a circuit is shown in Fig. 6-15.

Receivers for CW (Code)

Radio Amateur and many other receivers have provisions for reception of CW (code) signals.

These signals are produced at the transmitter by simply turning the transmitter carrier on and off in accordance with a code. For direct plain-language transmissions the code is the International Morse Code. Teletype transmission works similarly, except that often the carrier is shifted in frequency, rather than turned on and off. We shall discuss here the plain-language type of reception.

Because it is the otherwise unmodulated carrier that is turned on and off, there must be a BFO to make the signal audible at the receiver. Thus, a CW receiver must use a product detector. This is the only essential requirement for receiving CW signals.

Although not a necessity, it is considered a very important additional feature that a CW receiver have additional selectivity compared to those for voice and music reception. Because the only modulation of a CW signal is the relatively low-frequency on-and-off keying, this signal occupies a very narrow slot in the frequency spectrum. Thus, if the receiver has a very narrow pass band, it can separate received signals very close together in frequency, conserving valuable frequency spectrum. Various CW filters (installed in the IF amplifier) are available. Typical bandwidths are from 50 to 1000 Hz.

In a majority of cases, CW operation is an option included in receivers which are also usable for SSB and, sometimes, also AM. Since the filter pass band used for SSB must be wider than for CW to include the voice components of the sideband (usually about 2 to 3 kHz), there are usually two selectable filters furnished, one for SSB and one for CW. If the receiver is also designed for AM

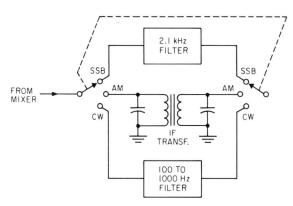

Fig. 6-16 *IF circuit in which selectivity can be switched to the desired amount for SSB, AM, or CW.*

reception, a third option is elimination of the filters and use of an IF transformer instead. Such an arrangement is illustrated in Fig. 6-16.

To add to the IF selectivity, CW receivers sometimes include (or are supplemented by external add-ons containing) AF filters. These provide a narrow AF pass band so that, in the AF amplifier section, the tone from the receiver can be selected from the tones of other received signals. Such a filter is shown in Fig. 6-17. Audio filters are often not limited to use for CW reception; some are also equipped to provide a low-pass characteristic (cutoff about 3 kHz) to minimize noise in voice communication receivers.

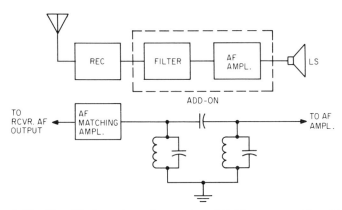

Fig. 6-17 *Simplified version of a passive AF filter to provide added selectivity for CW reception.*

Passive filters such as that in Fig. 6-17 provide no gain and are not usually variable in frequency. Active filters can provide both features. An example of an active filter circuit using an IC opamp is shown in Fig. 6-18.

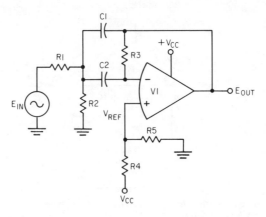

Fig. 6-18   *Active bandpass filter.*

Double-Conversion Receivers

Communications receivers most often operate in the HF (3–30 MHz), VHF (30–300 MHz), or UHF (300–3000 MHz) range. At frequencies above about 10 MHz, the problem of image response becomes important enough in single-conversion receivers to warrant special measures to control it. The image response frequency (at which such response is encountered) is:

$$f_{IR} = f_R \pm 2f_I$$

where

f_R = frequency of a given received signal to which the receiver is tuned

f_I = intermediate frequency

The plus sign applies when the oscillator frequency is higher, and the minus sign when it is lower, than the frequency of the received signal. The equation indicates that image interference is possible when the receiver is tuned to f_R and a strong interfering station comes on at $f_R \pm 2f_I$. Since both the received signal and the interfering signal out of the mixer are at the same intermediate frequency, they cannot be separated except in the antenna or RF amplifier section, where selectivity is more difficult than in the IF section. To ensure the desired degree of suppression of image response, the use of as high as possible an f_I is called for. Then, the spacing $2f_I$ becomes relatively great and good separation is much more easily attained. The only trouble is that, at a higher f_I, IF gain is much lower and good skirt selectivity (close to f_R) is harder to achieve.

For these reasons, many communications receivers use what is known as *double conversion.* In this system, the incoming signal is mixed with a first oscillator signal to beat it down to a moderate intermediate

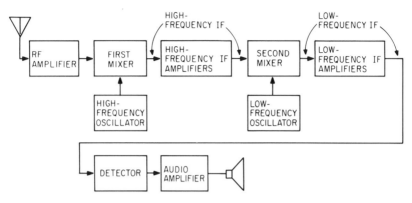

Fig. 6-19 *Block diagram of double-conversion superheterodyne receiver.*

frequency (say, from 2.5 to 10 MHz) first. This f_I is high enough so that the f_{IR} is easily rejected in the RF amplifier. Then, this moderate frequency IF signal is mixed with another oscillator signal to bring it down to or near 455 kHz, at which the high gain and skirt selectivity can be achieved. This is illustrated in Fig. 6-19. In Amateur Radio, consumer shortwave, and some communications receivers it is desirable to tune over a range of frequencies. However, often the first oscillator is crystal-controlled, and all stages up to the second mixer are made broad enough to tune over the full desired range of any particular band of frequencies (there are usually several switch-selected bands). A block diagram of such an arrangement is shown in Fig. 6-20. The RF amplifier and mixer are tuned by one ganged control over the entire frequency range of the receiver. Individual frequency bands are selected by making use of an appropriate crystal in the crystal oscillator. Tuning within any band is accomplished by tuning the calibrated dial of the second oscillator. This is a variable frequency oscillator

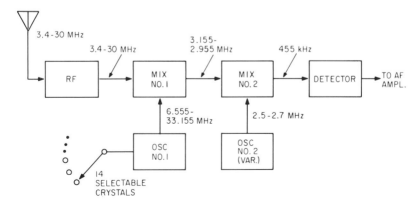

Fig. 6-20 *Frequency arrangement in an amateur communications receiver.*

(VFO) of good stability, which provides accurate selection of frequencies within any selected band.

Automatic Gain (Volume) Control

Automatic gain control (AGC), also called automatic volume control (AVC), uses a detector-derived dc signal as bias on the RF amplifier and/or the mixer and IF stages to control the amplitude of the signal to the last IF stage and the detector. If the amplitude of the received signal increases, the back-bias on the input stages increases, decreasing their gain and holding the output signal at a near constant level. This allows the full gain to be available for weak signals, while gain is reduced during reception of the strongest signals so that they do not overload the IF amplifiers or detector.

The idea of AGC is illustrated in Fig. 6-21(A), which is a block diagram of a receiver with AGC. The schematic diagram of the detector circuit and circuits that produce the AGC voltage are shown in Fig. 6-21(B).

The rectified dc voltage available across the detector load is used. It is filtered, then applied to the bases of the RF amplifier, mixer (converter), and the IF stages (some or all of these). The AGC voltage becomes part (in some cases all) of the bias voltage. If the signal tends to

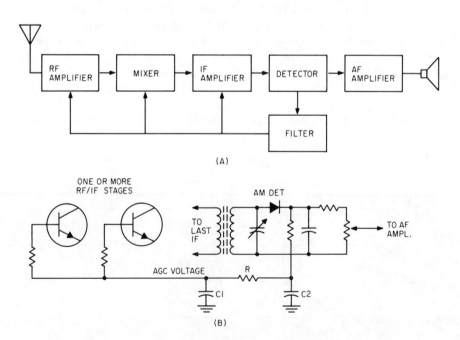

Fig. 6-21   AGC (AVC) in a superheterodyne receiver: (A) block diagram, and (B) schematic diagram of related portions.

be too great at the detector, it reverse-biases the amplifier, reduces gain, and thus keeps the output from increasing.

The example shown in Fig. 6-21(B) is for an AM receiver. The rectified signal in the detector is further filtered by R, C1, and C2 to remove the AF as well as RF fluctuations and produce a dc signal whose amplitude is proportional to the strength of the received signal. This voltage is applied to the IF amplifier stages and sometimes also to the RF amplifier and/or the mixer as bias. The greater the signal, the greater the reverse bias, which reduces gain to compensate.

Automatic Frequency Control

Another type of control widely used in FM and TV receivers is *automatic frequency control* (AFC). It employs a dc voltage, also derived from the detector, whose level is proportional to the deviation of the IF signal frequency from its proper value. This voltage is fed back so as to adjust the local oscillator frequency until the IF signal frequency is correct. This control is especially needed at VHF and UHF, where frequency drift is more of a problem than in the AM broadcast and HF ranges. The block diagram of Fig. 6-22 shows how AFC is accomplished in a typical FM receiver. Also, AFC is used in TV receivers to control vertical and horizontal deflection signal frequencies and automatic tint control (ATC). These are covered in Chapter 8.

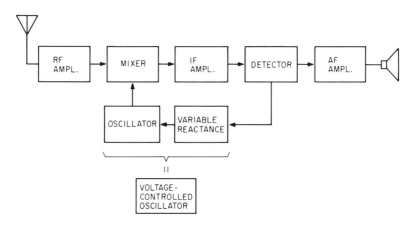

Fig. 6-22   AFC system block diagram.

The control voltage for AFC comes from an FM-type detector circuit, sometimes called a phase detector. This is the same as the Foster-Seeley detector illustrated in Fig. 6-10(B). The manner in which the AFC voltage is tapped off at the AF output point is shown in that figure. Filter circuit R3-C7 smoothes the AF fluctuations from the signal, leaving only the average dc output. If the discriminator is tuned precisely to the intermediate frequency and the received IF signal is centered at that frequency, the *average* instantaneous value of the AF output (and thus the AFC voltage) is zero. If the received IF signal center frequency drifts, the AF average and the AFC voltage become a positive or negative dc voltage, the polarity depending on which way the signal drifted.

The local oscillator in such a system is usually what is known as a *voltage-controlled oscillator* (VCO). It is an otherwise conventional oscillator set up with a varactor to which is applied the control voltage. The varactor exhibits a reactance whose value depends on the control voltage applied. The addition or subtraction of the reactance, as control voltage changes, corrects the tendency of the oscillator frequency to shift.

CHAPTER

7

Radio Transmitter and Transceiver Diagrams

Radio transmitters can be considered to include all devices used for transmission of intelligence by means of electromagnetic wave radiation. These include transmitters for broadcast stations (FM, AM, and TV), and communications (police, firefighting, commercial service, military, telephone, teletype, facsimile, CB, amateur, and so forth).

It is obviously not possible to go into all the diagrams of such a wide range of equipment here. Therefore, we shall discuss only representative examples of both continuous-wave (CW, for code transmission) and radiotelephone (voice and music) transmitters.

Radio transmitters use the principle that electric currents present at radio frequency in a suitable radiator (antenna) produce an electromagnetic field that radiates into space and is intercepted at a distant point and used by a radio receiver. Thus, the basic requirement of a transmitter is that it generate an RF current, called the *carrier*.

The carrier alone, although its presence can be detected at the distant receiver, cannot bear intelligence to the receiver unless it is *modulated*. The earliest method used for this was simply to turn the carrier on and off in a pattern corresponding to a code by which information can be transmitted. This is what happens in a code (or CW) transmitter. A device called a key is used by the operator to turn the carrier on and off in the right manner to convey intelligence.

For transmissions of voice and music, the sound must be converted to an electric current, which in turn causes variations in some feature of the carrier so that the variations can be reconverted to the original sounds at the receiving end.

Thus, in interpreting a schematic diagram of a transmitter, we must look for two main features: first, the generation of a radio frequency current (signal) and, second, some means of applying intelligence to (modulating) that signal. Thus, we can divide the diagram into an RF section and a modulator section. Of course, some stages are involved with both, as we shall see. In the case

133

of the CW (code) transmitter, the modulator may be nothing more than a means of connecting the key into the RF circuit. For sound transmission, however, the modulator becomes an appreciable part of the overall transmitter circuit.

The RF Section

As indicated above, the RF section must generate an RF carrier signal. Unless the transmitter is to have only very low power, the RF section must also include amplification to bring the carrier power up to the level required. There are transmitters of many different power output levels. The output level for a small "handy-talkie" may be less than one watt. CB transmitters are limited to 5 W (input to the last stage), and Amateur transmitters cannot exceed 1000 W maximum. AM broadcast transmitters in the US, if approved, can use up to 50,000 W (50 kW); some international shortwave broadcasting stations radiate hundreds of thousands of watts. Thus, the amount of RF amplification required depends upon what RF power output is required.

The RF section must also be designed to provide the means to select or adjust the frequency and to assure frequency stability. The frequency is controlled in the oscillator stage and it is there that stability is established and maintained. The operating frequency is determined by the oscillator frequency and any multiplier stages that follow.

Simple examples of RF section arrangements are shown in block diagram form in Fig. 7-1. The simplest RF section is just an oscillator; see Fig. 7-1 (A). This oscillator becomes a complete CW transmitter if turned on and off with a key. Such an arrangement is rare, because an oscillator alone cannot produce sufficient power while, at the same time, the frequency is kept stable.

The next stage in development of the RF section is adding a power

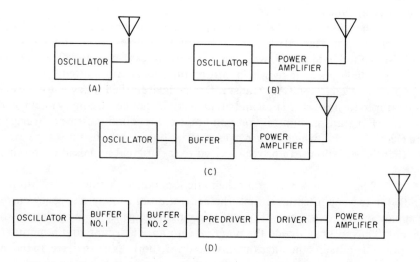

Fig. 7-1   *Transmitter RF section arrangements: (A) oscillator, (B) oscillator plus amplifier, (C) buffer added, and (D) with additional buffer and driver.*

amplifier; see Fig. 7-1(B). The oscillator needs to develop only enough power to drive the amplifier, which provides the higher output power required.

The arrangement in Fig. 7-1(C) goes a step further, providing more isolation of the oscillator from the antenna. An additional stage, called the *buffer*, is added between oscillator and amplifier. The increased isolation allows the oscillator to be almost completely immune to variations in antenna loading and amplifier adjustment.

It is not uncommon to encounter RF sections with more than one buffer and more than one power amplifier stage, as illustrated in Fig. 7-1(D). The additional buffers and drivers are more common today with solid-state devices than in previous times when tubes were used. As will be explained more fully later, tubes are still found in some final power amplifiers. We will now consider some typical circuits for the blocks in Fig. 7-1.

The Oscillator

Two general types of oscillator circuits are most widely used: (1) crystal-controlled oscillators, and (2) variable frequency oscillators (VFO). Crystal oscillator control makes use of a particular property of a quartz crystal wafer or slab. Each wafer has a physical resonance frequency at which it vibrates if properly excited. Because of a phenomenon called piezo electricity, the physical crystal vibration is translated into an electric voltage of the same frequency. Because of its physical rigidity, the crystal oscillates only (within close limits) at its natural resonance frequency and thus locks the oscillator frequency to it. A crystal-controlled oscillator has very high frequency stability. Only with great care in design can an oscillator without a crystal attain such stability.

A disadvantage of crystal control is that, unless the crystal is changed, operation is limited to the nominal crystal frequency or close to it. The frequency of a crystal-controlled oscillator can be changed a very small amount by changing external capacitance across the crystal, but such a change is limited to a small fraction of one percent of the nominal frequency. Provision is often made for a transmitter to operate on only one of a number of crystal-controlled frequencies by adding a switch to select the desired crystal from several incorporated into the design. But crystals alone are acceptable only where operation can be limited to a few set frequencies or small variations from any one of these frequencies.

Examples of crystal-controlled oscillator circuits are shown in Fig. 7-2. In Fig. 7-2(A) is a simple circuit that requires no coils (except the RF choke). It will be recognized as a Colpitts oscillator (see Fig. 2-44) in which the crystal substitutes for the coil. Sometimes the crystal is made part of the feedback path as in Fig. 7-2(B). This is basically a Hartley circuit (Fig. 2-43) that ordinarily oscillates at the resonance frequency of L1-C1. But, since the crystal is in the feedback path, it is obvious that feedback, and thus oscillation, can take place only at the series-resonance frequency of the crystal. Another version in which tuned circuit L1-C1 is not required is shown in Fig. 7-2(C). Notice that here crystal is in series with the feedback circuit from collector to base.

In many applications it is desirable to be able to change frequency

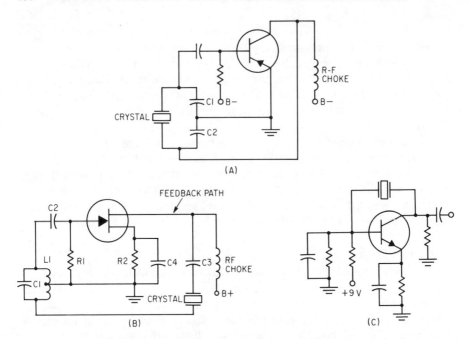

Fig. 7-2   *Crystal oscillator circuits: (A) Pierce, (B) Hartley, and (C) series feedback.*

smoothly from one value to another. All oscillators having variable frequency-determining elements (coil and/or capacitor) can do this if not locked into a crystal. Because the law requires a high degree of stability, however, careful design and construction are required to attain and maintain this stability without crystal control. Today, the term VFO is used to designate particularly those oscillators so designed. These are often referred to as *self-excited oscillators.*

Two examples of VFO circuits are shown in Fig. 7-3. In Fig. 7-3(A) is a Colpitts circuit using a bipolar transistor. Two variable capacitors are used: one for *bandsetting* and one for *bandspread* tuning. The former sets the general frequency range and the latter is for fine tuning. In Fig. 7-3(B) is a more elaborate verion of the same circuit principle, using a dual-gate MOSFET.

It should be emphasized that self-excited oscillators are subject to frequency change with a change of applied voltage. For this reason, in assessing a schematic diagram with a VFO in it, one should look for a power supply that is very stable, usually with voltage regulation, such as is discussed in Chap. 4.

In recognizing crystal oscillator circuits, the most obvious clue, of course, is the crystal itself. However, there are other nonoscillator circuits in some transmitters using crystals, so it is important also to note the position of the circuit in the overall schematic diagram. Labels are sufficient in most diagrams to identify the RF section and then from there one can trace back to the source. For VFOs, one must manage without the crystal, using the searching and tracing method.

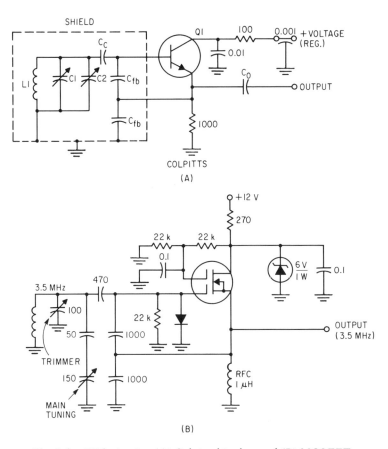

Fig. 7-3   *VFO circuits: (A) Colpitts bipolar, and (B) MOSFET.*

Buffer Stages

A buffer stage circuit has no particular distinguishing characteristics and must usually be recognized by its position in the overall circuit. There is often a buffer stage directly associated with the oscillator, to protect it from interaction. This buffer is usually operated class A to provide maximum isolation and seldom is tuned. An example of such a buffer circuit is shown in Fig. 7-4(A).

Other buffer circuits are tuned, as illustrated in Fig. 7-4(B). Whether a buffer stage is tuned or not depends partly on the design features of the whole transmitter. For example, some RF sections are *broadband*, that is, all the stages between the oscillator and the final amplifier are designed to operate over the entire tuning range of the oscillator, making it necessary only to gang the oscillator and final amplifier. Because a very broadly tuned amplifier stage has less gain than one sharply tuned, such a circuit may have more stages.

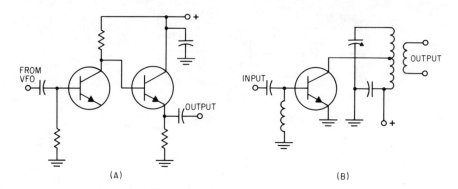

(A)   (B)

Fig. 7-4   *Buffer circuits: (A) untuned, and (B) tuned.*

Frequency Shifting by Multipliers and Mixers

For reasons of frequency stability, the RF oscillator in most transmitters is operated at a relatively low frequency (1 to 10 MHz). If the frequency of the RF output section is to be substantially higher, there are two methods available for making the necessary change between the oscillator and output frequencies: one is the use of *multiplier* stages; the other is the technique of *mixing* the oscillator signal with another signal.

Multiplier stages can be used for CW transmitters and for those phone transmitters in which amplitude modulation is not applied until after the multiplier stage. Multiplier stages cannot be used, however, with an amplitude modulated signal input because they introduce serious distortion. The multiplier operates on the principle that the nonlinearity of the circuit introduces considerable harmonic energy in the output circuit. The output circuit is tuned to the frequency of the desired harmonic. Harmonics fall at integral multiples; thus, if the input frequency is f, harmonics are at 2f, 3f, 4f, etc. The higher the multiple, the weaker the harmonic.

Figure 7-5 shows in block form how this might work out in specific numbers. An output frequency of 60 MHz is obtained from a 10-MHz crystal by use of a tripler and doubler.

As far as its schematic diagram is concerned, a basic, single-ended transmitter frequency multiplier does not differ from a straight-through amplifier. The biases may be adjusted a little differently to accentuate

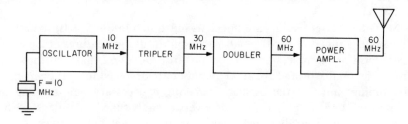

Fig. 7-5   *RF section using multiplier stages.*

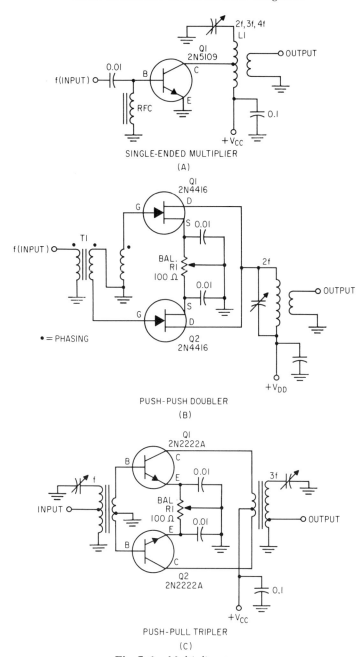

Fig. 7-6 *Multiplier stages.*

nonlinearity of the transfer characteristic and thus provide more harmonic energy in the output circuit.

Examples of some typical multiplier stages are shown in Fig. 7-6. In 7-6(A) is a single-transistor circuit which looks like a buffer, but its output circuit is tuned to the desired harmonic. In Fig. 7-6(B) and (C) are multiplier circuits using two transistors each. The *push-push* doubler of Fig. 7-6(B) has the advantage that the input signal frequency is cancelled in the two transistors. Similarly, the push-pull circuit of Fig. 7-6(C) is good for odd harmonics (3f, 5f, etc.) because the even harmonics (2f, 4f, etc.) are cancelled in the output circuit.

Although multipliers provide a good way to change the frequency of a signal from a low value to one much higher, the use of multipliers has two important limitations: the new frequency must be an exact multiple of the frequency of the source, and multipliers cannot be used with inputs that are already amplitude modulated. Since the latter rules out using a multiplier with the popular SSB type of signal, which is generated at low power and low frequency, it is an important limitation. The alternative is the use of *mixers*.

Mixers are generally familiar in their application in the superheterodyne receiver, in which they are used to mix an oscillator signal with the incoming signal, to obtain and IF signal, whose frequency is ordinarily lower than that of the received signal. It is lower because the difference signal ($f = f_{osc} - f_{input}$) is selected at the output of the mixer. The output signal frequency can be made higher by selecting the sum signal ($f = f_{osc} + f_{input}$) which is also available at the output of the mixer. The output frequency can be made any value desired by proper choice of oscillator frequency.

In receivers we have seen that trouble can be experienced in mixing if the intermediate frequency is relatively low, because it is then hard to exclude a signal on the image frequency; see Fig. 7-7(A). In transmitters, difficulties can be experienced if the frequency of one of the two mixed signals is very low relative to that of the other mixed signal and to that of the desired output signal. In such a case, the *sum beat* and *difference beat* outputs are so close in frequency that it is difficult to separate them.

An example is shown in Fig. 7-7(B). Signals having frequencies of 20 MHz and 0.5 MHz are mixed. The sum beat is at 20.5 MHz and the difference beat at 19.5 MHz. Some of the input signal at 20 MHz also gets through. If only

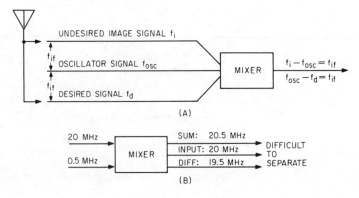

Fig. 7-7 *Mixing: (A) in receivers, and (B) in transmitters.*

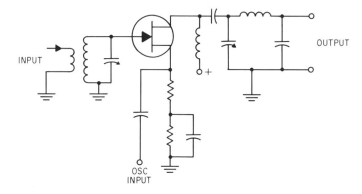

Fig. 7-8 *Mixer circuit.*

the 20.5-MHz sum beat is desired, it is difficult to provide sufficient selectivity in the output circuit to provide rather complete exclusion of the 20- and 19.5-MHz signals. For this reason, the mixing circuits encountered in practice keep the difference between the two input signals from being too great. Various arrangements are used in transmitter mixers. An example of a low-level mixer is shown in Fig. 7-8.

Drivers and Power Amplifiers

The final active device in the RF section is the power amplifier, which is usually preceded by one or more driver stages. Both power amplifier and driver stages differ from buffer stages only in bias adjustment and the fact that push-pull arrangements are most commonly used. A power amplifier is recognizable, even if it's unlabeled, by the fact that it is coupled to the antenna. A typical driver/power amplifier circuit is shown in Fig. 7-9. Another identifying feature to look for is a low-pass filter and/or standing wave meter between the power amplifier output and the antenna.

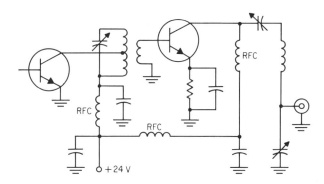

Fig. 7-9 *Driver/PA circuit.*

In most new equipment, virtually all vacuum tubes have been replaced by solid-state devices. In the RF power amplifiers of transmitters, however, use of vacuum tubes persists, probably because lower-voltage, high-current power supplies required for transistors are more expensive. It is therefore appropriate to consider here those symbols used for vacuum tubes and typical circuits for high power RF amplifiers employing tubes. The symbols are shown in Fig. 7-10. Also shown is an analogy between a bipolar transistor and a triode vacuum tube. Many circuits using vacuum tubes are very much the same as those using transistors, with the corresponding elements substituted as indicated in the

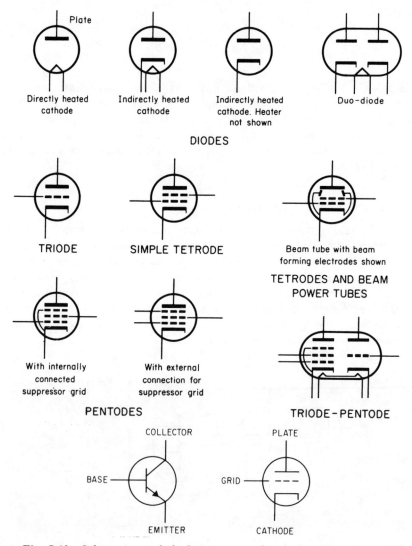

Fig. 7-10 *Schematic symbols for vacuum tubes and analogy between transistor and tube.*

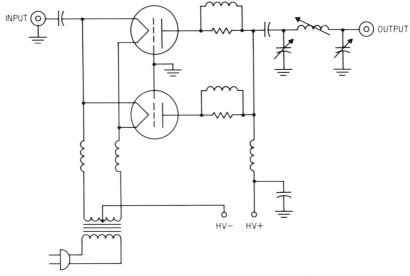

(A) TRIODES IN PARALLEL (GROUNDED GRIDS)

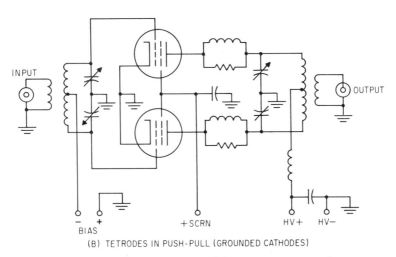

(B) TETRODES IN PUSH-PULL (GROUNDED CATHODES)

Fig. 7-11 *Typical RF final amplifiers using vacuum tubes.*

analogy. Of course, component values are different as a result of the differences in impedances and other factors. For example, a common-emitter circuit is like a common-cathode circuit and a common-base circuit is like a grounded-grid circuit.

Examples of final RF amplifier stage circuits using vacuum tubes are given in Fig. 7-11. Notice that the heater symbols are not always shown. In Fig. 7-11(A), the heaters act as the cathode, and thus their circuit is included. In Fig. 7-11(B), since tubes with separately heated cathodes are used, the heater circuit is omitted from the diagram.

Keying CW Transmitters

The RF section is basic to all transmitters because it provides a carrier. The simplest way to add intelligence is to key the carrier, that is, turn it on and off in accordance with a code, which for radio communication is normally the International Morse code.

Today, most code (CW) transmitters key in the earlier stages and the larger number keys the whole RF section, so that the transmitter operator can hear the signal from the operator at the other end, because the transmitter is off completely between the dots and dashes he is transmitting. If the local transmitter is not completely off when the key is up, the unkeyed transmitter output interferes with the local receiver. Keying the whole transmitter allows the other operator to break in while the local operator is sending.

If the oscillator is keyed, the whole transmitter is keyed, because the other stages are biased so that when they do not receive drive from the oscillator, they do not function. An example of a keyed-crystal oscillator circuit is shown in Fig. 7-12.

Sometimes a buffer stage may be keyed. A typical circuit for this is shown in Fig. 7-13. Of course, in this case, the oscillator and any preceding buffer stages stay on and break-in operation is not possible. Notice the keyer transistor used. When the key is closed, the transistor is biased into conduction to turn the buffer amplifier on.

Amplitude Modulation with Carrier and Sidebands

If voice or music (or, in some cases data) is to be transmitted, the carrier from the RF section must be *modulated* by that intelligence. In general, two types of modulation are universally used: *amplitude modulation* (AM) and *frequency modulation* (FM). Other systems, such as pulse-code and pulse-time modulation apply the pulses to the carrier by either amplitude or frequency modulation.

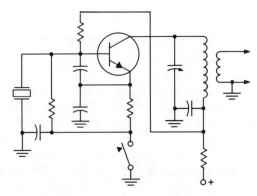

Fig. 7-12   Keyed crystal oscillator.

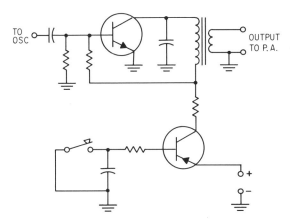

Fig. 7-13   Keyed buffer using keying transistor.

Amplitude modulation is used in any of several forms: carrier plus sidebands, single-sideband suppressed carrier, and double-sideband suppressed carrier. All are basically AM, since they are derived by varying the amplitude of the carrier. In practice, however, there has been a tendency to refer loosely to the first as AM, the second as single-sideband (SSB), and the third as double-sideband (DSB). Since the use of DSB is rare, we shall consider here only AM and SSB.

Amplitude modulation is most often accomplished by varying the collector (or, in a tube, the plate) current in accordance with the instantaneous variations of the AF modulating signal. The AF signal must be amplified so that there is sufficient power to cause the desired amount of modulation. The maximum AM feasible (without undue distortion or frequency bandwidth), that is, 100 percent, requires an AF power output equal to one-half the input power to the final amplifier to be modulated.

Two common methods for providing AM are shown in the diagrams of Fig. 7-14. In Fig. 7-14(A) is the method in which the output of the modulator is coupled to the final RF amplifier through a transformer. The coupling can also be accomplished with the use of an AF choke coil, as shown in Fig. 7-14(B). This is known as the Heising method.

The diagrams of Fig. 7-14 show modulation being applied to the final amplifier of the RF section. This is the most efficient arrangement as far as RF power is concerned. To reduce the AF power required, it is possible to modulate a preceding stage, which runs at lower power. When this is done, however, the final amplifier (or any other stages that follow the modulated stage) must be run as a linear amplifier, rather than in class C, which is more efficient, but which would cause distortion. A comparison of the two alternatives is shown in Fig. 7-15(A) and (B). Amplitude modulation is widely used in CB radios and is the standard modulation for the AM broadcast band.

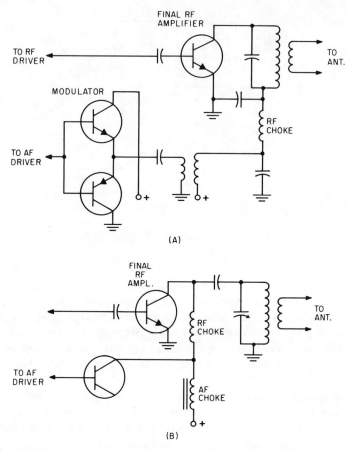

Fig. 7-14 *Simplified circuits for accomplishing amplitude modulation: (A) transformer-coupled, and (B) choke-coupled (Heising).*

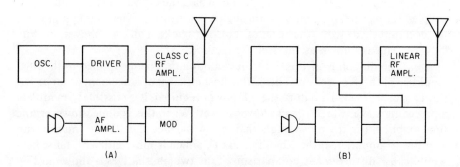

Fig. 7-15 *Comparison of (A) PA (high-level) modulator and (B) low-level modulator.*

Amplitude Modulation with Single-Sidebands

Because of its efficiency and other advantages, SSB transmission is becoming ever more popular and important. Because the transmitted signal contains almost no carrier, all of its power is concentrated in the sidebands, which contain the intelligence. This compares with 100 percent AM with full carrier and sidebands where two-thirds of the power is in the carrier, which does not carry the information. Thus, only one-sixth of the power is in each sideband. But, in SSB, if all the power available is concentrated in a single sideband, there is a great improvement in efficiency. At the same time, the frequency bandwidth is cut in half, because room is not needed for the other sideband.

Two methods have been used for generating an SSB signal: the filter and phasing methods. Since the phasing method is now a lot less common, we shall concentrate here on the filter method. In either case, the process starts by generating a full AM signal, then removing the carrier and one of the sidebands.

Functionally, the process is as shown in Fig. 7-16(A). An oscillator supplies an RF signal to a modulator, where it is modulated by the AF signal. The resulting full AM signal is applied to a device for cancelling out the carrier. The carrierless sidebands are then passed into a sharp filter, which eliminates one of them and produces an SSB signal.

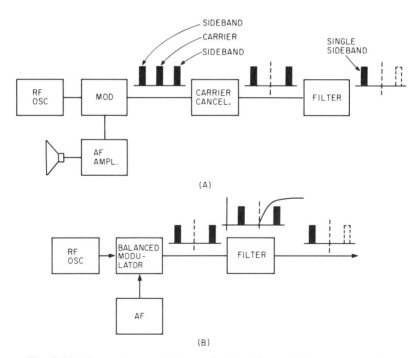

Fig. 7-16 *Generating an SSB signal: (A) with modulator and cancelling device, and (B) with balanced modulator.*

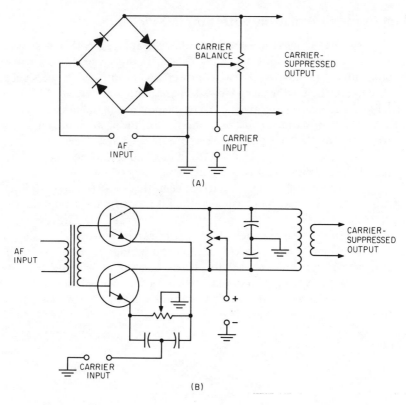

Fig. 7-17   *Balanced modulators: (A) with diodes, and (B) with transistors.*

In practice, the modulation and carrier cancellation take place in the same circuit, called a *balanced modulator*—Fig. 7-16(B)—which puts out a carrier-suppressed signal. A balanced modulator is simply a double-ended mixer in which the opposing two ends balance out the carrier. Balanced modulators use diodes or transistors. Examples of typical circuits are shown in Fig. 7-17. After the balanced modulator, the carrierless sidebands are passed through a filter circuit that cuts off one of them. This filter is similar to those discussed in Chap. 6 for receivers.

Many SSB transmitters provide a choice of transmission using either the upper sideband or the lower sideband. Theoretically, a simple way to do this is to have two filters, one to cut off each sideband, and be able to select either one. Since filters are costly, the choice is usually made by selecting one or the other of two crystals controlling the frequency of the carrier signal to the modulator. One places the signal higher than the filter center frequency, so that the lower sideband is filtered out and the upper sideband is transmitted. The other provides a lower frequency so that the lower sideband is transmitted and the upper is filtered out.

To attain proper stability and to make filtering more efficient, modulation and filtering are done at a low frequency. The standard intermediate frequency of 455 kHz is commonly used, so that in transceivers the IF amplifier can be used for both transmitting and receiving. Two important requirements result from the use of such a low frequency for SSB generation:

1. The suppressed carrier frequency must be raised considerably to a value usually in the HF region (3 to 30 MHz). The frequency cannot be increased by multipliers, since the modulated signal would distort. Therefore, it is done by mixing (heterodyning) the carrier signal with higher frequency signals.
2. Class C amplifiers cannot be used to amplify the signal; instead, lower efficiency class B or AB linear amplifiers are needed to avoid distortion.

As far as schematic diagrams are concerned, the class B and AB linear amplifiers do not appear any different from those operating class C. The differences exist only in the bias and drive applied. Thus, the RF section for SSB is similar to that of the same section for AM or CW, except that there may be one or more mixer sections instead of multiplier stages.

Frequency and Phase Modulation

Up to this point we have talked only about AM transmitters. The other major modulation type is a variation in carrier frequency (or phase). As indicated by its designation, in this method the *frequency* (or the phase) of the carrier is varied in accordance with the audio modulation. Whenever frequency modulation (FM) takes place, phase modulation (PM) also occurs, and vice versa. In frequency modulation, the carrier deviation varies only with modulating signal amplitude; in phase modulation, the frequency deviation varies with both amplitude and frequency of the modulation. Otherwise, they are the same. For this reason, an FM signal is often produced by phase modulating the carrier with an AF signal in which amplitude is inversely proportional to frequency.

Because it is the carrier frequency that is modulated, and not the amplitude, FM signals can be amplified in high-efficiency class C amplifiers without distortion. They can also be passed through multipliers, although these will multiply the deviation by the same ratio as the frequency. In some cases this is an advantage, because the deviation then required at the oscillator is relatively small. Otherwise, except for the modulator portion, an FM transmitter is very similar to a CW transmitter.

Modulation can be achieved either by *direct FM* or by phase modulation with an AF signal whose frequency characteristic is adjusted to compensate for the natural emphasis PM gives to the higher audio frequencies.

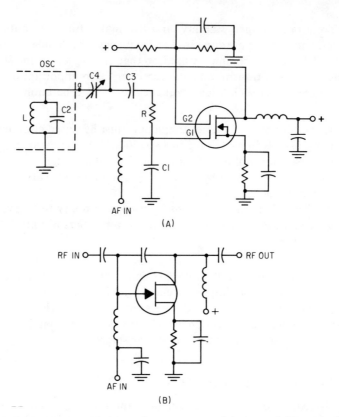

Fig. 7-18   *Direct FM: (A) with reactance modulator, and (B), with phase-shift modulator.*

Direct FM is usually accomplished by a *reactance modulator.* An example of this type of circuit is illustrated in Fig. 7-18(A). A particular feature which identifies the circuit is the combination of R and C1, connected across RF oscillator tuned circuit L-C2. The resistance of R is much higher than the reactance of C1, so that the total impedance is about equal to R. The RF current from the oscillator through the combination forms a voltage drop across C1 that lags the current by 90°. This drop, applied to gate one of the transistor, is amplified and the signal from the drain is then applied through C4 to the oscillator tuned circuit. This means that the circuit to the right of point *a* looks like a reactance to the tuned circuit and thus influences its resonance frequency. The AF modulating signal, also applied to G1, causes the gain of the transistor, and thus the magnitude of the reactance it exhibits to the oscillator tank, to vary in accord with the instantaneous voltage of that signal. Thus, the oscillator frequency is modulated.

   If the reactance modulator is applied to a crystal oscillator, very little frequency deviation takes place, because of the strong frequency control of the crystal. The same is true when modulation is applied to an amplifier, because the frequency is determined by an oscillator preceding the amplifier.

In order to optimize the frequency stability, frequency modulation is often accomplished by phase-modulating a crystal oscillator or an amplifier stage. As mentioned before, phase modulation is the same as frequency modulation if the AF signal frequency distribution is adjusted so that amplitude is inversely proportional to frequency. An example of a circuit for accomplishing phase modulation is shown in Fig. 7-18(B).

Broadcasts on the FM broadcast band (88–108 MHz) occupy a rather wide band because of the large deviation: ±75 kHz. This affords a significant advantage over AM noise and a wide modulation frequency range (to 15 kHz) for high-fidelity reception. FM is also widely used for communication, such as for police and fire departments, forest rangers, ham radio, etc., where a narrow band (±5 kHz) is used.

In FM broadcasting, signals are transmitted with preemphasis, an arrangement to minimize noise in reception. Much of the noise that may be experienced in reception is spread throughout the AF spectrum. But, the effect of the noise is reduced by decreasing the relative response of the receiver in the higher frequency portion of the AF spectrum, where noise content is greatest. To make this possible without loss of desired signal in that range, the higher frequency components of the AF signal are applied to the modulator in the transmitter at higher levels than the lower frequencies. At the receiver, the higher frequencies are attenuated, bringing the signal back to normal while minimizing noise. The emphasis at the transmitter is called *pre-emphasis* and the attenuation at the receiver is called *de-emphasis*. The simplest circuit for de-emphasis is shown in Fig. 7-19. The standard broadcast pre-emphasis is 75 μs, that is, the product of the resistance and capacitance in each circuit should be 75×10^{-6}. Exact values of separate resistors and capacitors to meet this are not often encountered, because the effects of other elements in the circuit must also be considered.

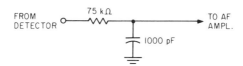

Fig. 7-19   *De-emphasis circuit.*

Transceivers

In communications work, it has been found convenient in many cases to combine the receiver and transmitter into one unit, called a *transceiver*. The combination offers not only the convenience of physical unity, but also the efficiency resulting from the fact that parts of the combination can be shared by both the transmitter and receiver portions.

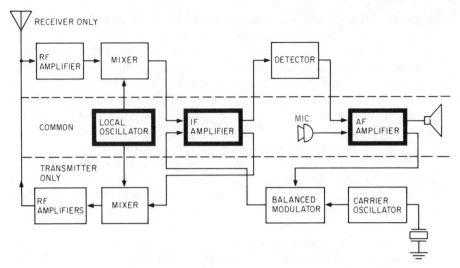

Fig. 7-20   *Simplified block diagram of transceiver.*

The individual circuit sections involved in a transceiver are for the most part the same as those already described for separate receivers and transmitters. Therefore, we shall discuss transceivers here only briefly, on a block diagram level.

A simplified block diagram of an SSB transceiver is shown in Fig. 7-20. Different arrangements are used in practice, but the one shown is typical. The sections used only for receiving are shown at the top, and those only for transmitting at the bottom. Those sections involved in both functions appear in the middle (enclosed in heavy lines).

Let us now trace the signal through the transmitter section. The carrier oscillator in the lower right-hand corner generates the transmitter RF signal at a relatively low frequency (usually 455 kHz, the intermediate frequency of the receiver). This signal is applied to the balanced modulator, which also receives the AF modulation signal from the AF amplifier. The balanced modulator combines these two and produces a DSB, suppressed-carrier signal. This signal goes to the 455-kHz IF amplifier. The IF amplifier contains a filter, which eliminates the unwanted sideband; it also amplifies the resulting SSB signal. The IF output is applied to the transmitter mixer, where it is mixed with the local oscillator signal to produce a signal of the desired radiated signal frequency. This signal is then amplified and applied to the antenna for radiation. The local oscillator is tunable and is used to adjust the carrier to the desired transmission frequency.

Now consider the receiver. The received signal comes from the antenna into the RF amplifier, where it is amplified and applied to the receiver mixer, which combines it with the local oscillator signal to bring its frequency down to 455 kHz, the intermediate frequency. The received signal is then fed to the IF amplifier, which provides the high selectivity of the filter and amplifies the signal. Then the IF signal goes to the detector, which for SSB is usually a product detector. The recovered AF signal is then amplified in the AF amplifier and reproduced by the loudspeaker.

One important feature of a transceiver is the fact that the operating frequencies of the transmitter and the receiver are the same. This is because the same local oscillator converts the transmitter's generated 455-kHz signal up to the transmitted signal frequency and the received signal down to the 455-kHz receiver intermediate frequency. The transceiver's calibrated tuning dial controls this oscillator and thus the common operating frequency. This is important because, for most communication circuits, the transmitters at both ends operate on the same frequency.

8

Interpreting Television Receiver Diagrams

Probably the most complex piece of equipment most people will encounter is the modern TV receiver. Consequently, the schematic diagram for a TV receiver, as might be expected, is large and complex. For those who are familiar with radio receivers, the layout of a TV receiver is considerably clarified. A TV receiver is essentially a radio receiver. The same kinds of radio waves are used for transmission, propagation, and reception. The fact that a picture must be transmitted and received, as well as sound, simply means that a somewhat different pattern of modulation (on an extra carrier) must be used for the picture. The portion of the TV system for sound is almost exactly like that for the FM broadcast system. The picture portion is like an AM broadcast system, except that the modulation is picture signals, known as *video*, plus control pulses.

Because of the size and complexity of the schematic diagram of a TV receiver, we shall look at the overall layout from only the block diagram standpoint. Then, we shall examine the schematic diagrams of some of its significant parts.

Block Diagram of Complete Receiver

The most widely used TV set is the color type, although many black-and-white (BW) sets are still in use. Because a color TV receiver can be considered as a combination of a basic BW receiver with sections to handle color added, we shall discuss the BW portions first, then proceed to those used only for color.

As in the case of examples given in previous chapters, the location and context of any given circuit in the whole scheme of things are important in the interpretation of the schematic diagrams of any parts of the circuit. For this reason, we start with a brief review of the circuit sections.

The block diagram of the BW portion of a color TV receiver is shown in Fig. 8-1. This could also be considered the block diagram of a BW TV receiver,

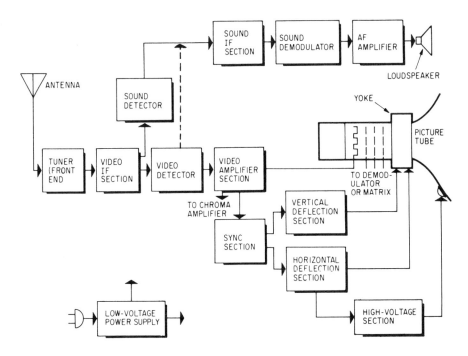

Fig. 8-1 *Block diagram of the BW portion of a color TV receiver.*

the only difference being that, in the BW receiver, the separate *sound detector* is not used; the video detector there doubles for both functions and the sound IF signal goes from the video detector to the sound IF amplifier, as indicated by the dashed line.

The signal from the antenna enters via the tuner or *front end,* which includes RF amplifier stage, oscillator, and mixer. The nature of the signal is shown in Fig. 8-2, which includes the frequency spectra of (A), the BW signal, and (B) the color signal. Important to notice is the fact that there are two separate carriers: one for sound and one for picture. Also observe that the bandwidth of the signal, including both sound and picture sidebands, is nearly 6 MHz (which is the frequency space allowed for each TV channel).

Returning to the block diagram, notice that all of the frequency components must be handled in the tuner and video IF amplifier sections. After this, the amplified signals are divided between the sound detector and the video detector.

The output of the mixer contains two useful IF signals: the video IF signal and the sound IF signal, 4.5 MHz lower in frequency than the video IF signal. The video IF amplifier is designed to amplify both the sound IF and the video IF signals, and it applies both of them to the video detector and also to the sound detector. In the sound detector they mix (heterodyne) to produce a third signal, which has a carrier frequency of 4.5 MHz, the difference between the carrier frequencies of the two IF signals. The 4.5-MHz carrier is frequency-

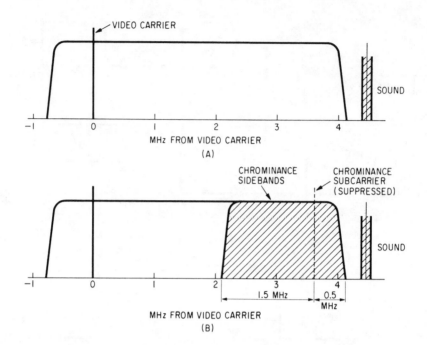

Fig. 8-2   *Spectrum of TV signals: (A) BW, and (B) color.*

modulated by the audio signal and, in addition, is amplitude-modulated to some extent by the video signal. The amplitude modulation is removed by the limiting action of the succeeding sound stage. The sound IF signal of 4.5 MHz is amplified in the sound IF amplifier and demodulated in the sound demodulator. The resulting AF signal is then amplified in the AF amplifier and applied to the loudspeaker. Virtually all modern receivers use at least one IC in the sound section.

The output of the video IF amplifier section also goes to the video detector, where the video components of the signal are separated. The components include those for a BW picture (luminance information), those for adding color (chrominance information) and those needed to synchronize the picture and its color (sync information).

Figure 8-1 shows the destinations of those signals needed for BW reception. The video signal goes to the video amplifier, and then to the grid of the picture tube. The picture sync components go the sync section, where they are processed to control the vertical and horizontal sweep oscillators and to supply pulses for the high-voltage power supply (and sometimes for part of the low-voltage supply).

The additional sections needed for color reception are indicated in Fig. 8-3. Notice that a delay line is added in the path of the BW (luminance) portion of the signal. This delays the luminance signal enough to compensate for

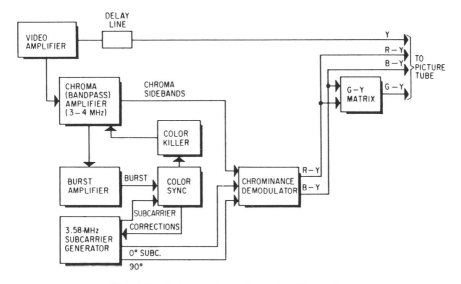

Fig. 8-3 *Color sections of a color TV receiver.*

the delays the color signals undergo in the extra sections they must traverse, so that all the signals arrive at the picture tube at the same time.

The other output of the video amplifier is filtered and amplified in the *chroma bandpass* amplifier. The chrominance elements go to the chrominance demodulator, and from there they go to the picture tube. The demodulation process separates the two chrominance signals, R-Y and B-Y, which are combined in proper proportions to produce the G-Y signal. These three difference signals are then applied to the picture tube (or, in some cases, matrixed to produce R, G, and B for the tube).

Another signal from the chroma amplifier goes to the *burst amplifier*, which separates the color burst signal from the sync pulses. This signal is used to synchronize the two subcarriers used in the chrominance demodulator, so that the color components are in proper sync. The color killer turns off the chrominance amplifier if no color signals are being received.

Tuners

Functionally, the first circuit of a TV receiver is the *tuner*, which includes the RF amplifier, the mixer, and the oscillator. It is usually packaged as a separate module and can be replaced as a unit if necessary.

The TV channel frequency spectrum has two main portions: VHF, including channels 2 through 13 (54 to 216 MHz), and UHF, channels 14 through 83 (470 to 890 MHz). While there is a trend toward a single tuner covering both ranges, separate VHF and UHF tuners are still found in most receivers. A popular arrangement is that shown by the block diagram in

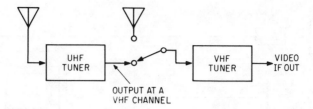

Fig. 8-4   *Diagram showing how the UHF tuner output is fed to the VHF tuner.*

Fig. 8-4. The VHF tuner is used as the "first IF" section for the UHF tuner that precedes it. This "first IF" is the frequency band of a locally unused VHF channel to which the VHF tuner is set for UHF reception.

Because of the large number of channels which the tuner must cover, one of the outstanding recognizable characteristics of a tuner is the switching arrangement. Switching of some kind is usually found in a VHF tuner, but sometimes a continuous (rather than detented) tuning mechanism is used for UHF.

Figure 8-6 (opposite page) is a typical tuner with a multisection switch for channel selection. Notice that four coils are switched: RF amplifier input and output, mixer input, and oscillator. Conventional common-emitter transistor amplifier and mixer circuits are used and the IF output is taken from transformer T2.

A number of tuners now use varactor switching instead of mechanical-type RF switches. The principle is illustrated in the simplified schematic diagram of Fig. 8-5. A varactor is a special kind of diode that has the characteristic of a capacitor when a dc voltage is applied across it. The magnitude of the capacitance is controlled by the amount of voltage applied. If the varactor

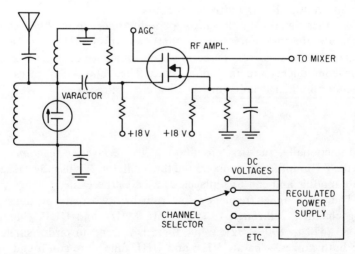

Fig. 8-5   *The principle of varactor tuning, as applied to the input of a TV RF amplifier.*

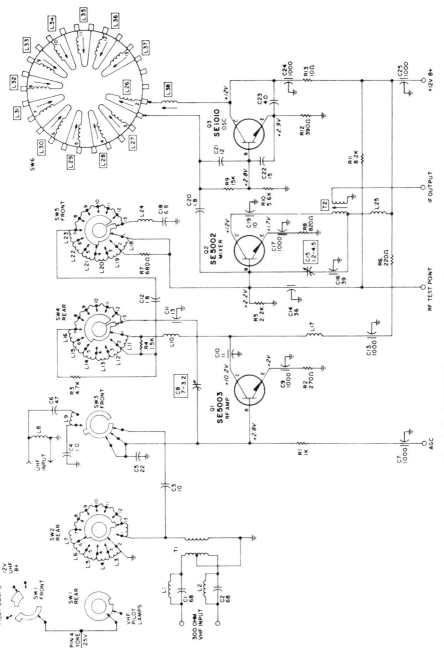

Fig. 8-6 *VHF TV tuner, using multisection switch.*

becomes part of a tuned circuit, its capacitance tunes the resonant frequency. By adjustment of the dc voltage on the varactor, one can then tune for resonant circuit to any desired frequency within its range. In Fig. 8-5, the varactor is part of the RF amplifier tuned input circuit. The dc voltages are available from a regulated power supply; each voltage is adjusted to tune the circuit to a desired channel. A channel switch is used to select the channel wanted, and similar circuits are included to tune the mixer and the oscillator. Although this system requires switching, it is only a dc circuit, and no RF switching is necessary.

PLL Synthesizers

Another system used for channel selection in some tuners is the *phase-locked loop* (PLL) synthesizer system. PLL is a system whereby a signal of any of a wide range of frequencies can be frequency-stabilized by control of one standard reference oscillator.

The operating principles of a PLL synthesizer are illustrated in Fig. 8-7. The voltage-controlled oscillator (VCO) circuit contains a varactor, which varies the oscillator frequency in accordance with the magnitude of a dc voltage applied to the varactor. The reference oscillator is usually a fixed-frequency, crystal-controlled circuit whose accuracy is reliable. In the instance shown, the reference oscillator frequency is relatively low, so that its stability is ensured. Outputs from the VCO and reference oscillator are applied to a phase comparator. However, because the VCO frequency is much higher than that of the reference oscillator, the VCO signal is passed through a divider, which reduces its frequency to a value close to that of the reference oscillator.

The phases and frequencies of the two signals are compared in the comparator, and, if they differ, a dc voltage develops. This difference voltage is fed back to the VCO to correct the oscillator frequency until its divided frequency again matches that of the reference.

Because programmed dividers can operate at any integral multiple of the oscillator frequency, the setting of *n* sets the desired oscillator frequency and

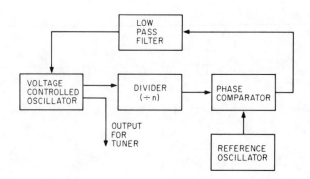

Fig. 8-7   *Principle of a phase-locked-loop (PLL) synthesizer.*

the channel to be received. PLL synthesizer setups are found in only a few TV receivers, but are most frequently encountered in CB transceivers.

Video IF Amplifier and Detector

From the standpoint of their schematics, TV video IF amplifiers are similar to those in radio receivers. Examples are shown in Chap. 6. Functionally, the TV IF must amplify a wide band of frequencies (nearly 6 MHz) to include the carriers and sidebands indicated in Fig. 8-2. The whole video IF amplifier in any given receiver must have a frequency response that maintains the ratio of the video IF amplitude to sound IF amplitude. A typical response is shown in Fig. 8-8.

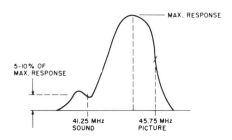

Fig. 8-8   *Typical TV IF response curve.*

Since most recent TV receivers use ICs in the IF amplifier, these will be encountered in a schematic diagram of that portion. A typical IF section is shown in Fig. 8-9. Notice that a single IC is used for IF and detector sections and is shown in three parts (IC1A, IC1B, and IC1C) and that the tuned circuits are external to them. The IC contains the active components (transistors, etc.). IF alignment is accomplished by means of the slugs in the IF transformers, which is the reason for keeping them physically accessible. The diagram also shows the video detector and the circuit of the first video amplifier.

Notice the coupling transformer between IC3 and the video amplifier, which acts as a trap, to eliminate the 4.5 MHz sound IF signal from the video and sync composite signals. The 4.5 MHz sound IF signal is taken from the output of IC2, as indicated. In color receivers, there is a separate detector for sound. It separates the 4.5 MHz sound IF signal from the remainder of the video IF signal. As mentioned before, this signal is the result of mixing the high-frequency sound IF signal (41.25 MHz) and video IF signal (45.75 MHz), and its frequency is the difference between them.

Video Amplifier

After the video signal is retrieved by the video detector from the IF signal, it is amplified, usually, by several stages of amplification. The video

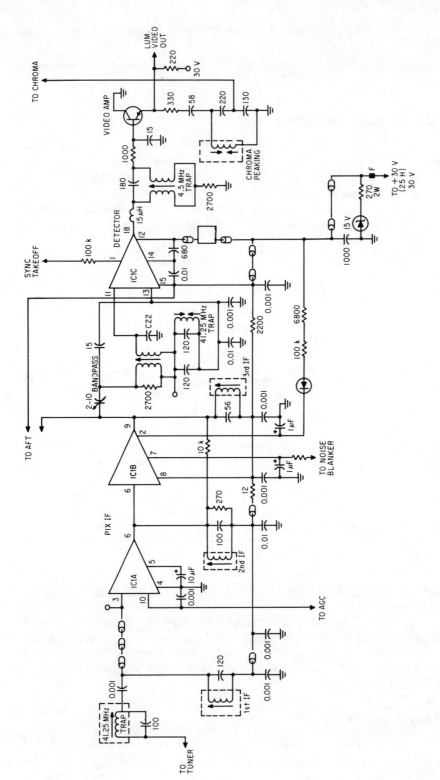

Fig. 8-9 *Typical video IF section.*

amplifier supplies signals to three other receiver sections: (1) the picture tube (through matrices), (2) the chrominance (also known as *chroma* or *bandpass*) amplifier, and (3) the picture sync section. The video signal that goes to the picture tube is the *luminance* or Y signal; it contains all the video signal components for a monochrome or BW picture. (The signals to the chrominance amplifier and sync section are considered within the next few pages.) The luminance signal goes to the picture tube matrixing circuit, where it is mixed with chrominance signals to produce the signals applied to the picture tube grids and cathode. A typical video amplifier stage circuit is shown in Fig. 8-10. There are usually at least two stages of video amplification.

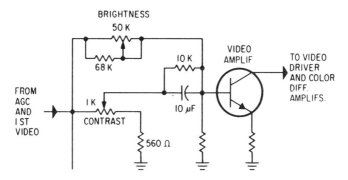

Fig. 8-10 *Video amplifier stage circuit.*

The Sound Section

Before following the video section further, we shall backtrack just a little, to trace the path of the sound signal. A block diagram of a typical sound section is shown in Fig. 8-11.

As previously indicated, the sound and video carriers in the video IF signal are 4.5 MHz apart. The IF signal (containing both signals) is fed to a

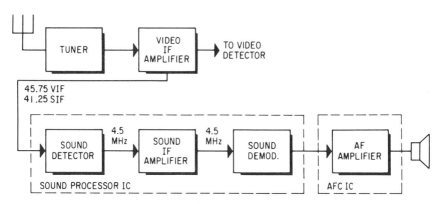

Fig. 8-11 *Block diagram of TV receiver sound section.*

separate detector, in which the two signals are heterodyned. The difference heterodyne, based at 4.5 MHz, is selected and used as the sound IF signal. In the detector, the high-frequency sound IF signal is kept at a lower level than the video IF signal, so that the FM sound modulation is predominant in the 4.5-MHz sound IF signal.

This 4.5-MHz IF signal is then amplified and applied to the FM demodulator, such as the discriminator or ratio detector described in Chap. 6. The resulting AF signal is amplified in a rather conventional amplifier similar to those discussed in Chap. 5.

In practically all modern TV receivers, the AF section is either all or partly included in an IC. A typical division into two ICs is shown by dashed lines in Fig. 8-11. In Fig. 8-12 is an example of a circuit in which all of the AF components and the sound IF amplifier are included in one IC. Figure 8-12(A) shows the three main functions accomplished in the single IC illustrated schematically in Fig. 8-12(B). In Fig. 8-13, only the sound IF amplifier and demodulator are included in the IC. The AF stages that follow use discrete components.

Picture Sync and Deflection

Included in the received video signal are sync pulses, which are used to keep the vertical and horizontal deflection signals synchronized to the received

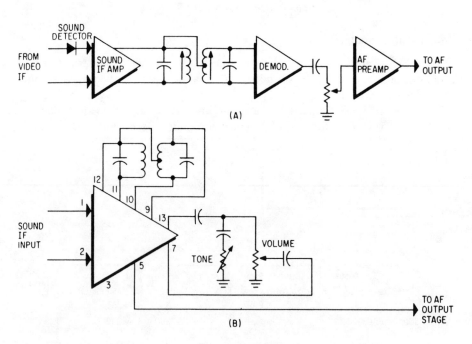

Fig. 8-12 *Typical IC sound section: (A) breakdown of major parts of the IC, and (B) circuit using the overall symbol for the IC.*

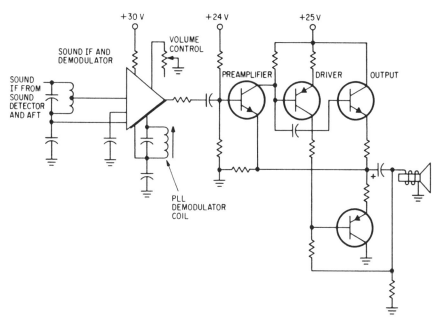

Fig. 8-13 *Sound section in which only the sound IF and demodulator are included in the IC, with discrete transistors being used elsewhere.*

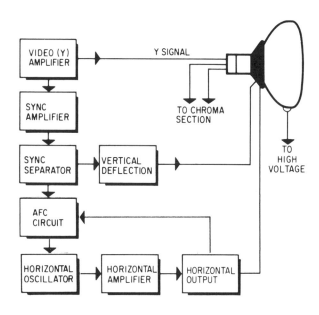

Fig. 8-14 *Picture sync and deflection section in block form.*

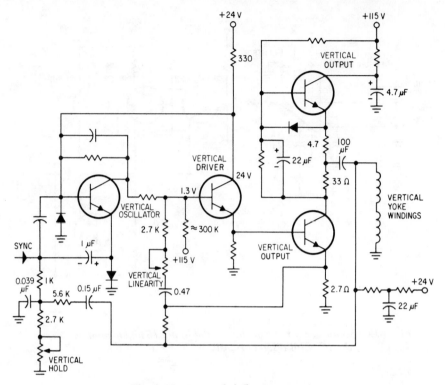

Fig. 8-15   *Vertical deflection circuit.*

signal. Sync pulses are separated from the video signal and are divided into vertical and horizontal pulses. Sample pulses from the receiver's deflection circuits are compared with the received pulses, which lock the received picture into frequency and phase.

A block diagram of the picture sync and deflection circuits appears in Fig. 8-14. Notice that the separated vertical sync pulses go directly to the vertical deflection section, where they precisely control the frequency of the vertical oscillator. The horizontal sync pulses are applied to the automatic frequency control (AFC) circuit, where they are compared to the pulses from the horizontal deflection circuit. Any difference results in a correction voltage which is fed back from the AFC circuit to the oscillator. As indicated, the deflection signal sample used for comparison in the AFC is often taken from the output stage of the horizontal deflection section.

A typical vertical deflection circuit is shown in Fig. 8-15. The sync pulses are applied to the oscillator base circuit where they trigger each cycle and keep the sweep in sync with the picture. The oscillator feeds an emitter-coupled stage which drives a series-type output amplifier similar to those frequently used in AF amplifiers. This output goes to the vertical deflection yoke windings to sweep the picture tube beam.

A typical horizontal deflection circuit is shown in Fig. 8-16. In this case, the oscillator is synchronized by a dc voltage from the AFC circuit. This voltage

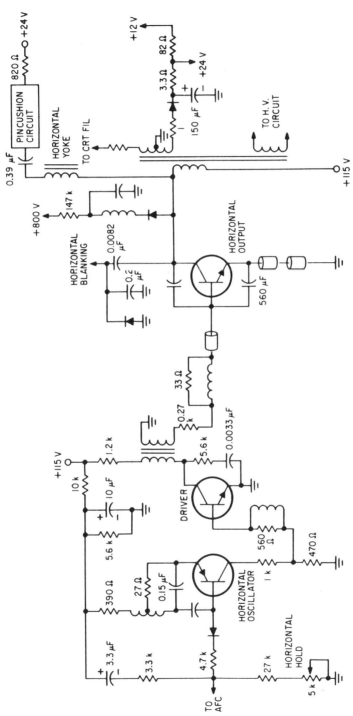

Fig. 8-16 Horizontal deflection circuit using transistor output.

corrects any deviations of oscillator frequency and phase sensed in the comparison in AFC. A single power-type transistor is used in the output stage. As indicated, besides its basic purpose of providing deflection current, this circuit also produces horizontal blanking (to cut off the beam of the picture tube during retrace), current for pincushion magnets (to adjust for minimum picture distortion at the edges of the screen), heater current for the picture tube, pulses for the high-voltage power supply, and a source for some of the low-voltage supply. The circuit for the low-voltage (12-V and 24-V) supply is included in the figure. The high-voltage power supply is discussed further in this chapter.

A variation of the output circuit, in which silicon controlled rectifiers (SCRs) instead of transistor(s) are used, is now popular. An example of such a circuit is shown in Fig. 8-17. The inductance of the yoke windings and capacitances C2 and C3 form resonant circuits. The yoke, with C2 in series with C3, oscillates at about 35 kHz, and with C3 alone at about 5 kHz. The SCRs and diodes switch these components in such a way as to use part of a 5-kHz oscillation for the longer trace period and part of a 35-kHz oscillation for the rapid retrace interval.

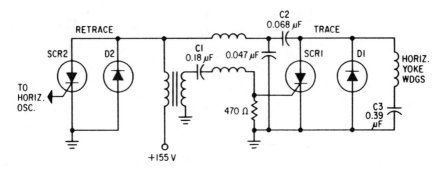

Fig. 8-17 *Circuit showing the use of SCRs in the horizontal output section.*

Power Supplies

There are two power supplies in a TV receiver: low-voltage and high-voltage. Low-voltage supplies provide current to transistors and other active devices requiring voltages in the 12- to 200-V range, and to the screen grids of the picture tube. The high-voltage supply is for the voltage (in the order of 20 to 30 kV) for the anode and the focus anode of the picture tube (5 to 15 kV). Some low-voltage (LV) supplies and all high-voltage (HV) supplies get their input from the horizontal output stage. Since this current is at a frequency of 15,734 Hz, filtering is much simpler than with 60-Hz supplies. However, some kind of 60-HZ supply must always be used, because current is needed to operate the horizontal deflection section, to provide energy for the 15-kHz supplies.

The typical configuration of the 60-Hz type of LV power supply is shown in Fig. 8-18. In this example, a transformer with two separate secondary

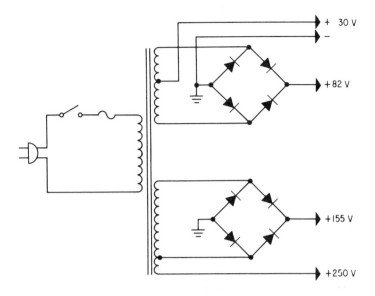

Fig. 8-18 *60-Hz type low-voltage power supply.*

windings, one of which is tapped, is used to provide the variety of voltages required. In each of the outputs is a conventional filter circuit to smooth the dc output.

An example of the other type of LV supply, operating from the horizontal output circuit, is shown in Fig. 8-19. In this case, the ac voltages are taken from the transformer at the input of the horizontal output amplifier, although often they are taken from its output. Notice the use of a Zener diode regulator for the 20-V output.

High-voltage power supplies always derive their power from the horizontal output stage. One winding (or an extension of another) on the horizontal output transformer provides high-voltage pulses which are rectified or passed into a voltage multiplier which steps up voltage while it rectifies the signal. A simple high-voltage power supply circuit using a voltage tripler is shown in Fig. 8-20. Focus voltage, needed for the picture tube, is also obtained from this supply.

Chrominance (Bandpass) Amplifier

The signal from the video amplifier is also supplied to the chrominance (chroma) amplifier. This amplifier forms an active high-pass filter that separates the chrominance sidebands from the remainder of the video signal. (The frequency arrangement of these signals is shown in Fig. 8-2.) The chrominance sidebands are part of the video signal, but constitute in themselves a modulated signal that must be demodulated to recover the chrominance signals.

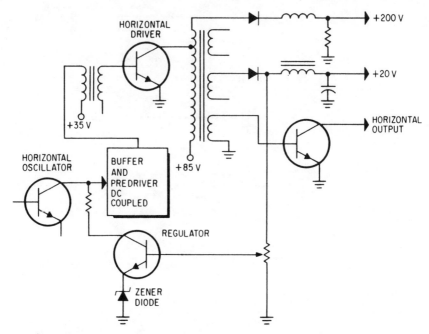

Fig. 8-19   *Low-voltage power supply receiving its input from the horizontal deflection section.*

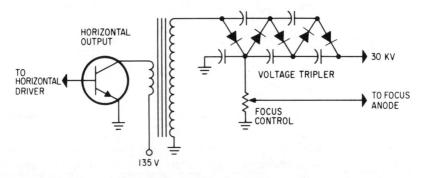

Fig. 8-20   *High-voltage power supply.*

It can be seen in Fig. 8-2(B) that the chrominance sidebands occupy the upper portion of the video IF spectrum, from about 3 MHz upward. For proper demodulation and use of these sidebands, they must be separated from the luminance video (Y signal) occupying mainly the range below 3 MHz. This separation is accomplished by the chrominance amplifier, and this is why it is often referred to as the *bandpass* amplifier. As far as the circuit is concerned, it is not much different from the video amplifier. The circuit constants are chosen, however, to limit response to the 1 MHz between 3 and 4 MHz. A typical chrominance amplifier is shown in Fig. 8-21.

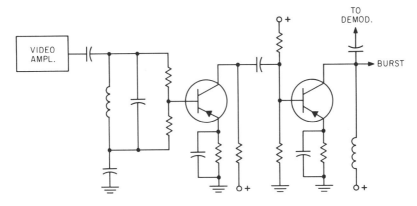

Fig. 8-21 *Circuit of a chrominance amplifier stage.*

Color Sync

The demodulation of the chrominance sidebands requires the addition of subcarriers (two or three) to them to produce color signals of a form suitable for use at the picture tube. The phases of the subcarrier signals must be carefully controlled (in sync) in relation to a *burst signal* which is superimposed alongside each horizontal pulse. Thus, the color sync portion of the receiver separates the color burst signal from the video signal, generates its own subcarrier signal, locks it to the burst signal, and then applies the resulting signal to the demodulator.

The manner in which the color sync operation fits into the overall chrominance portion of the receiver is shown in Fig. 8-22(A). The demodulator in this case, calls for three subcarrier signals, one each to demodulate the G-Y, B-Y, and R-Y *color difference signals.* The subcarrier signals are really all from the same oscillator, but two of them are shifted in phase to produce the desired result.

The color burst is separated from the output of the chrominance amplifier by the burst amplifier. A typical circuit is shown in Fig. 8-23. The burst amplifier is turned on only during the horizontal pulse pedestal interval, when the color burst is received. It amplifies the burst signal and applies it to the phase detector. Also applied to the phase detector is the signal from the subcarrier generator, which is a crystal-controlled oscillator operating at 3.579545 MHz, which we refer to here as 3.58 MHz. Although the frequency of this oscillator is closely controlled by the crystal, it can still be pulled a few Hz one way or the other by a change of reactance in the circuit. The received burst signal and the signal from the crystal oscillator are compared in the phase detector. If the oscillator signal frequency or phase tends to deviate from that of the burst signal, the phase detector sends a dc voltage reflecting the change to the oscillator. The voltage is applied to a varactor in the oscillator circuit, changing its reactance to correct the tendency of the oscillator frequency to deviate.

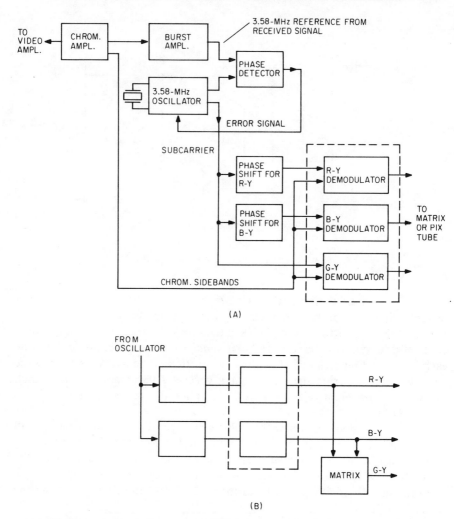

(A)

(B)

Fig. 8-22   *Color sync and demodulator arrangements: (A) separate G-Y demodulator, and (B) G-Y matrixed from R-Y and B-Y.*

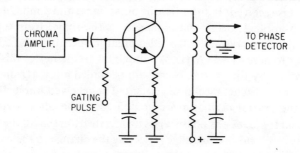

Fig. 8-23   *Diagram of a burst amplifier.*

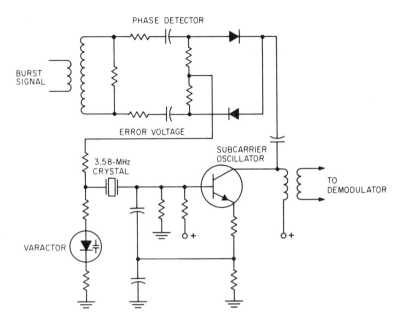

Fig. 8-24 *Subcarrier oscillator and color sync phase detector.*

A representative circuit for these functions is illustrated in Fig. 8-24. Notice the varactor in the input (base) circuit of the oscillator, where it acts as a capacitor. The *error voltage* from the phase detector is applied to it, to change its capacitance as needed. Notice also the oscillator output signal coupled from the collector to the detector and the burst signal going to the input of the detector at the left.

Thus, the color sync section generates the subcarrier for use in the demodulators and synchronizes it (subcarrier) to the incoming burst signal. The subcarrier is then ready for application to the demodulator.

Color Demodulation

Turn again to Fig. 8-22(A) and notice that the same subcarrier signal is applied to each demodulator unit. The difference between the demodulator units is the phase of the subcarrier signal it uses. The same chrominance sidebands are demodulated into color signals of different content, depending on the subcarrier phase. The desired color outputs are color-difference signals R-Y, B-Y, and G-Y. In the system shown in Fig. 8-22(A), the signal directly from the oscillator demodulates G-Y. The different phases needed for R-Y and B-Y are obtained by phase-shift circuits. Quite often, only the R-Y and B-Y signals are derived, and G-Y is obtained by mixing these two in proper proportions in a matrix. This arrangement is illustrated in Fig. 8-22(B).

A wide range of options is available by using different subcarrier phases, but most fall into the categories just discussed. One pair of color-difference signals sometimes used are referred to as X and Z color signals. They are chosen to minimize the chrominance signal bandwidth required. Older receivers demodulated to I and Q color signals; these required nearly 3.5 MHz instead of the present 1 MHz between 3 and 4 MHz in the video spectrum.

A typical simplified color demodulator circuit of the type depicted in Fig. 8-22(A) is shown in Fig. 8-25. The circuits of the demodulators are virtually identical to that of a phase detector.

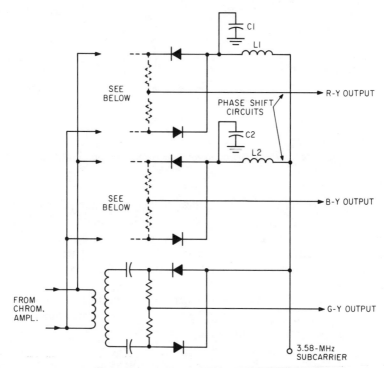

Fig. 8-25   *Color demodulator circuit.*

Application to Picture Tube

Once the color (difference) and Y signals are available, they must be combined and fed to the picture tube. The most common type of picture tube has three grids, each with its corresponding cathode. One each of these is for the R, G, and B signals. Thus, the difference signals must be combined with the Y signal so as to cancel out Y and come out with R, G, and B.

Several common arrangements for combining or applying color signals to the picture tube are shown in Fig. 8-26. In Fig. 8-26(A) the three difference

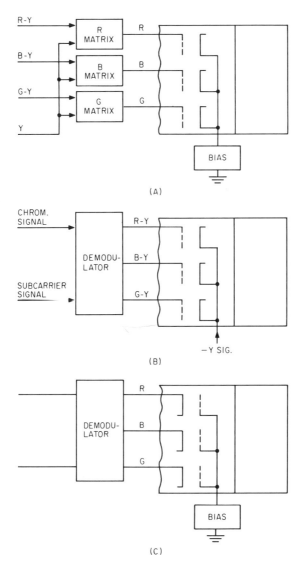

Fig. 8-26 *Three ways of applying color signals to the picture tube: (A) difference signals matrixed to R, B, and G before the tube; (B) difference signals to the grids, with the Y signal applied to the cathodes; and (C) application of R, B, and G from the demodulator to the cathodes.*

signals are combined in matrices with the Y signal, to derive R, B and G. These signals are then, respectively, applied to the three grids. The cathodes have dc bias, but are grounded to the signals. In the arrangement of Fig. 8-26(B), difference signals are applied to the grids. A minus Y signal is applied to the cathodes. This causes the Y signal to add to each color-difference signal. The result is that, between the grid and the cathode of each gun in the tube, is applied one of the three signals, R, G, and B. The arrangement of Fig. 8-26(C) is the reverse of that of (A), in that the color signals are applied to the cathodes, and the grids are at signal ground.

A typical circuit for the color amplifier/matrix portion of a receiver is shown in Fig. 8-27. This shows the location of a brightness control and the three color drive controls which are used to adjust relative amplitudes of the three color signals. Also shown is a typical arrangement for adjusting the relative levels of the screen grid voltages of the picture tube.

ICs In the Chroma Section

The color sync/demodulator portion of the receiver is often referred to as the *color processor*. In most receivers some or all of this section is included in one or more ICs.

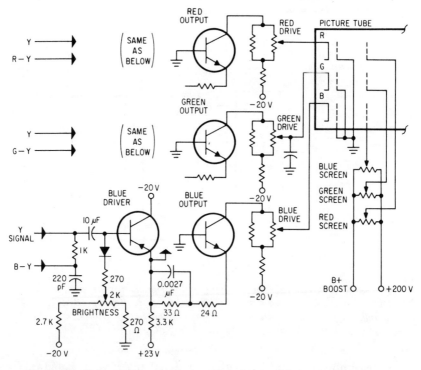

Fig. 8-27   *Color output circuit in which R, G, and B are applied to the cathodes of the picture tube.*

An example of a circuit using one IC for the demodulators and following color amplifiers is shown in Fig. 8-28. The demodulators are supplied from a voltage regulator also included in the IC.

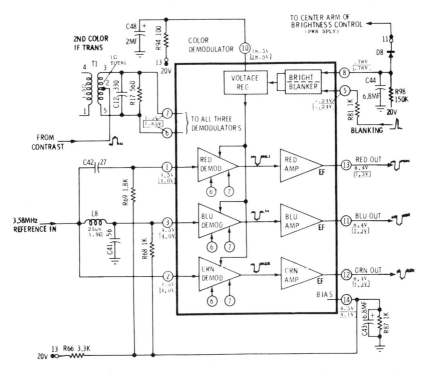

Fig. 8-28 *Circuit showing how color demodulators can be incorporated into one IC.*

Some Other Circuits

The complexity of complete TV receiver circuits does not allow consideration of every section, nor does interpretation of schematic diagrams in general require this. Instead, we shall now briefly consider, in block diagram form, the schemes used for a number of the automatic control circuits encountered in TV receivers.

Every TV receiver today has *automatic gain control* (AGC). This reduces receiver RF and IF gain during reception of strong signals, so that they will not overload the amplifiers. A block diagram of an AGC system is shown in Fig. 8-29. Because the video modulation level varies widely during any typical program, the sync pulse level alone is constant enough to be used as a measure of signal strength. This level is determined by separating the sync pulses by gating as indicated at the *gated AGC amplifier* block. The clamping circuit forms these pulses into a dc voltage whose value is representative of the received signal level. This voltage is used as bias on the tuner and IF amplifier. Its polarity is such that, as signal level increases, gain of the stages decreases, so that the signal strength at the output of the IF amplifier remains essentially stable.

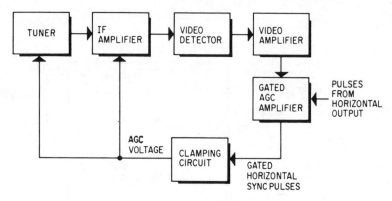

Fig. 8-29   *AGC system.*

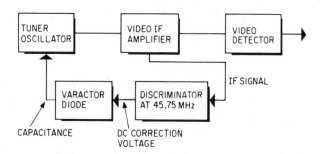

Fig. 8-30   *AFT system.*

To keep a TV receiver tuned properly on a station, a circuit called *automatic fine tuning* (AFT) is used. A block diagram is found in Fig. 8-30. An IF signal from the video IF amplifier is fed to a discriminator tuned to 45.75 MHz, the video intermediate frequency. If the frequency of the local oscillator in the tuner is such that the video IF signal is right at 45.75 MHz, where it belongs, the discriminator dc output voltage is zero and no adjustment is made at the tuner oscillator. If the tuner oscillator is off its proper frequency, the video IF signal is also incorrect. The result is that the discriminator puts out a dc error signal of a polarity depending on whether the tuner oscillator frequency is high or low. The dc error voltage is applied to the varactor, which accordingly changes the capacitance it applies to the tuner oscillator. This adjustment returns the oscillator frequency to its proper value.

CHAPTER

9
Computer
Diagrams

In Chap. 2, a number of the basic digital logic packages used in computers and other digital equipment were discussed. In this chapter, to illustrate computer diagramming, we show a few examples of how these basic packages are combined to make functional units of computers.

Overall Computer Organization

The basic units normally required in a computer are shown in the block diagram of Fig. 9-1. As shown here, the five basic sections are: arithmetic unit, memory, control unit, and input and output circuits.

Arithmetic unit (sometimes called the *arithmetic logic unit* (ALU)): This is where the actual basic arithmetic functions, such as adding, subtracting, multiplying, and dividing are accomplished.

Memory: This is where information is stored. Part of it is for the program, which is a series of instructions telling the other computer sections what to do in the proper order. The other part of the memory stores data, which may be fixed numbers used in calculations.

Control unit: This section controls the timing and functional relationships of all the different parts of the computer so that they may best respond to the program.

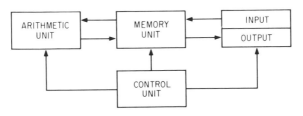

Fig. 9-1 *Simplified block diagram of a computer.*

Input and output circuits: These connect the computer with the outside world. The input information needed by the computer must be in a particular digital form not convenient or practical for the user to provide. The input circuit takes the user's information and converts it to the form needed for the sections within the computer. The output circuit does the same thing in reverse: it takes the output of the computer sections and converts them into data useful to the user.

Use of Binary Number System

The numbers we use in everyday life make up the decimal system. Decimal means having to do with the number 10, which means that the decimal number system comprises ten digital symbols: 0, 1, 2, 3, 4, 5, 6, 7, 8, and 9. Although there would be some advantage in using decimal numbers in computers, because of their familiarity, their use would complicate computer circuits. It is the basic nature of most computer circuits to have only two states, which can be characterized as *on* and *off.* Thus, a number system with only two digital symbols is best suited to these circuits. Such a system is called the *binary system.* It uses only 0s and 1s. The basic nature of the system is shown in Fig. 9-2.

A complete discussion of the binary system is not in order here. However, Fig. 9-2 should provide sufficient feel for the subject for interpretation of equipment diagrams. Figure 9-2(A) shows a list of decimal numbers 0 through 16, 32, and 100, with their equivalent binary numbers. A study of how the binary numbers develop will indicate how the system works. Because only two symbols are available, the numbers in binary are naturally much longer than their decimal counterparts.

Also, the diagram in Fig. 9-2(B) shows eight digital positions with the value that each represents. Notice that each digit position has a value of 2 to some power. Starting at the *least significant digit,* that is, the one at the right-hand end, the digit positions in the binary system represent successively 2^0, 2^1, 2^2, etc.—in this case up through 2^7. Each of these values is multiplied by the number digit in its position and added to the products obtained in all the other positions. Of course, in a binary number, the number to multiply by is never greater than 1, so for each digit, it is either the basic value or nothing. The figure shows how the number depicted is thus evaluated as a decimal number (165).

To show that the principles are the same for a decimal number, the same explanation of the same number in decimal form is also shown in Fig. 9-2(C). Because the decimal system is based on 10 symbols (0 through 9), it is 10 that is raised to the appropriate power for each digit. A number system can be based on any number. For example, evaluation of a "base 8" or octonary number is shown in Fig. 9-2(D).

In binary, it is common to refer to the digit positions in a number as *bits.* The left-most digit (which multiplies by the highest power of the base) is called the *most significant bit* and the right-most digit is called the *least significant bit.*

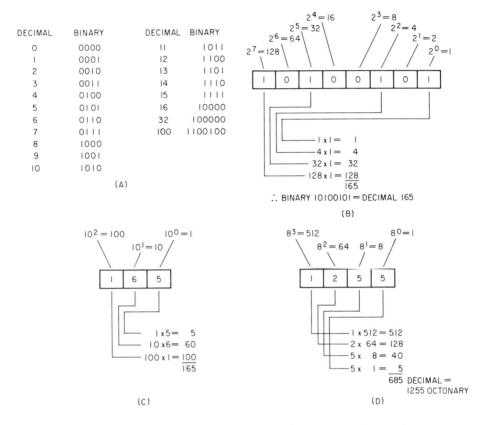

Fig. 9-2 *The binary number system: (A) binary equivalents of some decimal numbers, (B) evaluation of a binary number in terms of a decimal number, (C) the digits of the same decimal number following the same principle, and (D) evaluating an octal number in terms of decimal number.*

Arithmetic (Logic) Unit

As its name implies, the arithmetic unit accomplishes the purely arithmetic operations. It does these mostly with circuits called adders, subtractors (modified adders), shift registers, and volatile (short storage time) memory.

Probably the most representative unit in the arithmetic unit is the adder. An adder has two parts, each called a half adder. A simple half-adder circuit and its schematic symbol are shown in Fig. 9-3(A and B). It consists of an AND and an EXCLUSIVE OR, connected as shown. It operates on one bit (from each number to be added) at a time. Thus, the input at A and B can each be a 0 or a 1. Notice that the output includes not only a sum, but also a carry. This is because in binary, $1 + 1 = 10$, since 10 is the equivalent of decimal 2. Since there is no 2 in binary, a 1 must be *carried* over to the next leftward position. All the half-adder does is produce a single-digit sum and the carry. As we shall see, another half-adder must be included to complete the addition for two digit numbers, and more full adders allow more digits to be handled.

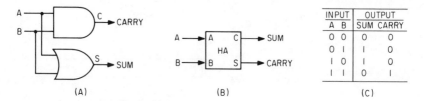

Fig. 9-3   *The half-adder: (A) circuit, (B) schematic symbol, and (C) truth table.*

Figure 9-3 (C) shows the truth table for the half-adder. Notice that only when inputs A and B are 1 is the carry a 1. The sum is then 0. This table can be verified by checking the signals through the AND and the EXCLUSIVE OR according to the principles discussed in Chap. 2 for the different combinations of inputs.

Figure 9-4 shows how the half-adder can be combined with another half-adder and an OR to form a full adder. It also shows the truth table for the combination. The additional feature here is that the full adder can handle a carry in from a previous adder as well as put the proper carry out. The truth table shows the results for each combination of inputs.

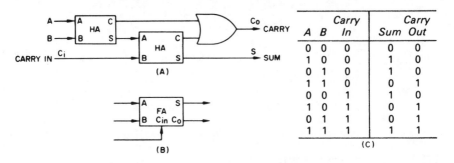

Fig. 9-4   *The full-adder: (A) circuit, (B) schematic symbol, and (C) truth table.*

Parallel and Series Adders

As indicated in the above discussion, an adder works on a single pair of bits (one from each number to be added) at a time. As soon as the bits arrive at its input, it operates and produces an answer. If, as is usual, numbers with quite a few digits are to be added, these digits must be ordered and applied to the adder or adders so that the answer comes out in the proper sequence of bits. This can be done in either of two ways: serially or in parallel. Parallel adders are now more popular, because they are faster, although they require more adder circuits.

The arrangement for a serial adder is shown in Fig. 9-5. The numbers to be added are fed to the input through devices called *registers*. These devices

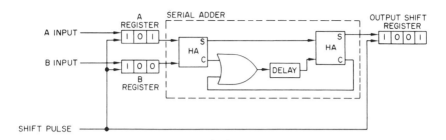

Fig. 9-5 *Serial adder unit.*

contain the two numbers and feed them out, digit by digit, so that corresponding digits arrive at the adder at the same time. Typically, a register is a series of flip-flops. If a digit is fed into the left-most flip-flop, it moves to the right one flip-flop each time the system is triggered. Thus, a whole number, set up in the register, can be moved from the left to the right one digit each time it is triggered, and the number "spills" out at the right-hand end, in this case to our adder. Thus, a register is a temporary storage from which a number can be "cranked out" upon command. The manner in which registers are shown schematically is illustrated in Fig. 9-5.

In a parallel adder arrangement, an adder for each digit of the numbers to be added is used, as illustrated in Fig. 9-6. Each adder must handle only one pair of digits, and thus the whole addition is accomplished during the time of a one-digit addition. Thus, the parallel adder must be used in high-speed computers. The serial adder, in contrast, must consume the time it takes for all the digit-pair additions to take place.

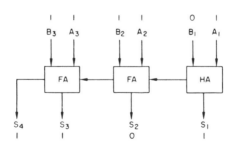

Fig. 9-6 *Parallel adder: the numbers indicate the addition of 111 (A) and 110 (B), for a sum of 1101.*

Subtractor

The subtractor is simply an adder which adds the two numbers after changing one to what is known as its "twos complement." Diagrams of a half-subtractor and a full-subtractor, along with schematic symbols and truth tables, are shown in Fig. 9-7 (A) and (B), respectively. Notice that the half-subtractor is

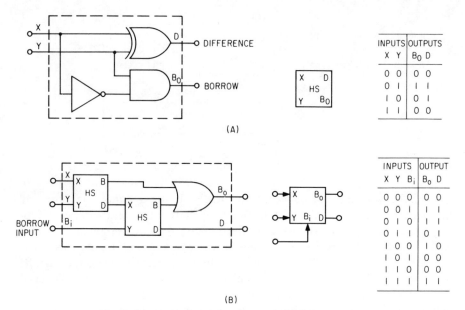

Fig. 9-7 *(A) Half-subtractor, and (B) full-subtractor.*

the same as the half-adder, except for the inclusion of an inverter, and the fact that the sum is now the difference and the carry has been replaced by the borrow.

Adder-Subtractor Unit

Figure 9-8 shows a diagram of a parallel binary adder-subtractor unit. This unit uses registers at both the input (incident register) and output (accumulate register) of the adders. The output number is fed into the accumulate register, where it can be kept until a write signal releases it through AND D. In this unit, the new number in the incident register is added to the number already in the accumulator register, thereby updating the accumulator. Notice that the input number is entered serially into the incident register, but this number is applied in parallel to the full-adders. Each bit of the number goes to a separate full-adder. Notice also that an *overflow circuit* is also provided. If the sum is too large to fit into the accumulator, this circuit produces an overflow indicator bit. The complement blocks complement the input number for subtraction, and the sign control block is to handle numbers of different polarities.

Timing Circuits

The circuits of a computer must be synchronized with each other. For example, for the adder we saw how the input words must appear at the input simultaneously (each bit) so that they can be properly added. All operations

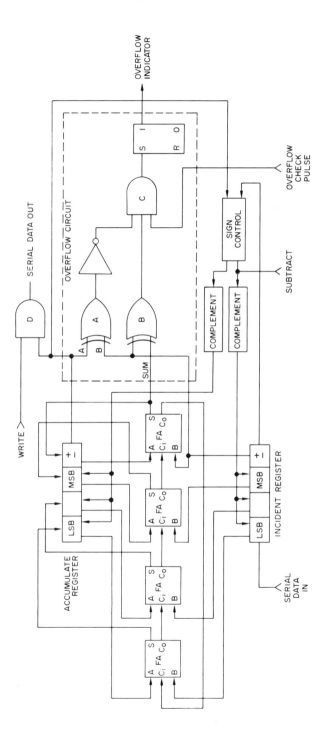

Fig. 9-8 *Basic parallel binary adder-subtractor unit.*

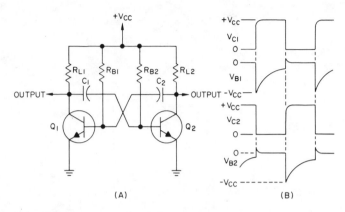

Fig. 9-9   *Free-running multivibrator (such as might be used to generate timing signals).*

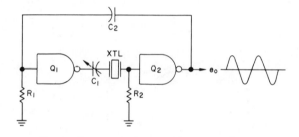

Fig. 9-10   *Crystal-controlled oscillator for timing.*

throughout the computer must be similarly synchronized. To provide for this, some kind of common timing system is used. Usually, a free-running oscillator serves this purpose, with its output properly shaped to trigger the digital circuits as required.

One type of timing-circuit oscillator is the free-running multivibrator, whose circuit is illustrated in Fig. 9-9. Another is a crystal-controlled oscillator, whose circuit is shown in Fig. 9-10.

Counters

Another circuit common in computers and other digital equipment is the *counter*. This is a device that indicates how many pulses enter it within a given period. Its operation is very similar to that of the register, which we have seen applied in the serial adder.

Figure 9-11 (A) shows the basic arrangement of a counter. It is a series of flip-flops. The first clock pulse applied to the left-most flip-flop trigger causes that flip-flop to change state so that Q changes from 0 to 1. The next pulse from the clock changes Q of the first flip-flop back to 0. Thus, as the

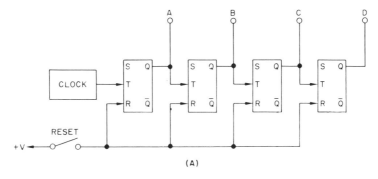

D	C	B	A	Clock Count
0	0	0	0	0
0	0	0	1	1
0	0	1	0	2
0	0	1	1	3
0	1	0	0	4
0	1	0	1	5
0	1	1	0	6
0	1	1	1	7
1	0	0	0	8
1	0	0	1	9
1	0	1	0	10
1	0	1	1	11
1	1	0	0	12
1	1	0	1	13
1	1	1	0	14
1	1	1	1	15
0	0	0	0	16

(B)

Fig. 9-11   *Binary counter using a series of flip-flops.*

uniform stream of pulses comes from the clock, Q of the first flip-flop becomes 1 every other clock pulse, with a zero between.

Each succeeding flip-flop is triggered when the output of the previous flip-flop goes from 1 to 0. The output of the first flip-flop does this once for each two input clock pulses. The third flip-flop has the same relationship to the second, so it is triggered every four clock pulses, and the fourth flip-flop is triggered every eight clock pulses.

If we form a truth table of the Q outputs related to the number of clock pulses (arranging the outputs in reverse order) we come out with the table shown in Fig. 9-11(B). Notice that the outputs form, in each case, a binary number equal to the number of pulses counted. Thus, we have a *binary counter.*

In the output of the counter, we might want to indicate the results in decimal numbers. A way this can be done is illustrated in Fig. 9-12. The Q and Q̄ outputs of the flip-flops are combined for the inputs of each AND so as to

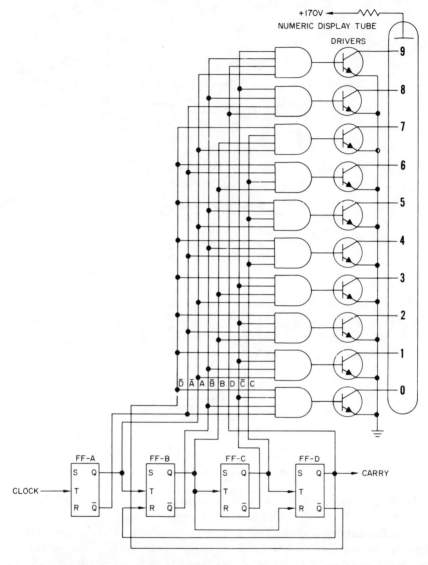

Fig. 9-12 *Decade counter with binary-to-decimal decoder and decimal number display.*

operate the AND only when the binary output number corresponds to the decimal digit fed by that AND. For example, take the decimal number 5, which corresponds to binary 0101. The AND for 5 is operated when Q of FF-A, $\overline{Q}$ of FF-B, Q of FF-C, and $\overline{Q}$ of FF-D are all 1. Checking the effects of five successive clock pulses shows this to be the correct condition.

Dividers

In connection with TV receivers, in Chap. 8, we discussed the phase-locked-loop (PLL) system, in which dividers are used. Frequency dividers take an ac or pulse signal and put out a signal with a frequency equal to the input frequency divided by some whole number. Checking the counter we have just considered, we can see that it is also a divider. By responding to every other clock pulse, the first flip-flop acts as a two-to-one frequency divider. Going through two flip-flops divides the frequency by four, three by eight, etc. This gives an idea of the variety of uses to which such a circuit can be put.

Conclusion

We have considered here only a few examples of the circuits used in computers and other digital devices. The purpose has been to provide some familiarity with how these circuits are presented in circuit diagrams. Obviously, there is not space in this chapter (or this book), nor is it its purpose, to provide any complete coverage of theory or to catalog circuits. For such information, the reader is referred to theory courses in electricity, electronics, and computers. In such cases, it is expected that a familiarity with schematic diagrams, developed through this book, will stand him in good stead.

Index

Accumulators, 184
Active devices, 76,77
Adders, 181-184,186
Adder-subtractor units, 184,185
Alternators, 9
AM (amplitude modulation), 97,101,111,
 114,116-118,120-123,126,127,131,138,
 144-149,151,154,156
Amperes, 12,74,75
Amplifiers, 23,24,37-41,86,101-103,
 134-137,145
 audio, 97-100,120,121
 burst, 157,171,172
 cascaded, 38
 chroma bandpass, 157,163,169-171
 common-collector, 44
 common-emitter, 44,108,114,158
 common-source, 39
 differential, 57,58,105,106,108
 FET, 40
 gain-controlled, 29
 IF (intermediate frequency), 38,116-118,
 122,126,128,149,152
 JFET, 39
 MOSFET, 39
 operational, 57-59,127
 phase inverter, 39-40
 power, 98,100,103,104,109,120,141,142
 push-pull, 39,120
 resistance coupled, 37
 RF (radio frequency), 15,23,112-116,
 122,128-130,141-143,152
 single-ended, 39
 transformer-coupled, 38
 video, 155-157,161-163,169-172,176-178
 voltage, 98,101,103-107,120
AND, 48-50,181,182,187
 table, 50
Anode, 24,31
Antennas, 111-114,116,133,135,141,152
Arrays
 diode, 56,57
 integrated circuit, 57
 transistor, 56,57
Audio systems, 65,97-110
Automatic fine tuning, 178
Automatic frequency control, 131,132,
 166,168

Automatic gain (volume) control, 113,
 118,122,130,131,177,178
Automatic tint control, 131
Automobile radios, 78
Autotransformers, 18
Avalanche breakdown, 30

Balance controls, 103
Base, 25,54
Batteries, 7,8,10,77,78,94
Battery chargers, 78,95,96
Bias, 100,113,130,131,138,141,144,176
 back, 130
 currents, 76
 forward, 37,89
 negative, 88
Bias polarity, transistor, 26,27
Binary system, 180,181
Binding posts, 7
Bits, 180
Block diagrams, 3,62-66,70
 amplifier, 116
 audio system, 101
 automatic frequency control, 131
 automatic gain (volume) control, 130
 battery charger, 96
 computer, 179
 microcomputer system, 66
 radio receiver, 63,64,66,112,129
 transceiver, 152
 TV receiver, 155,163,165
 voltage regulator, 90
Brushes, 9
Buffers, 134,135,137-139,141,144
Bus, 5

Cabling, 67
Capacitance, standard, 73,74
Capacitors, 12-15,35,37,76,85,86,92-94
 coupling, 99
 decoupling, 38
 dielectric, 12,13
 electrolytic, 13,14
 fixed, 13,14
 ganged, 13-15,113,114
 limiting, 120